The BDA Guide to
SUCCESSFUL BRICKWORK

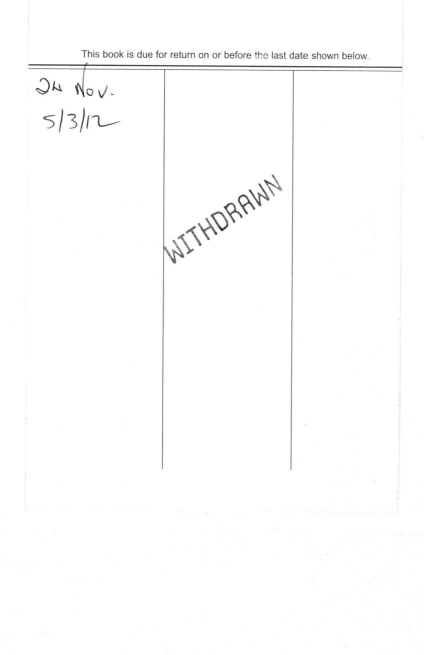

The BDA Guide to SUCCESSFUL BRICKWORK

Second Edition

THE
BRICK
DEVELOPMENT
ASSOCIATION

Woodside House, Winkfield, Windsor, Berks

ELSEVIER
BUTTERWORTH
HEINEMANN

AMSTERDAM • BOSTON • HEIDELBERG • LONDON • NEW YORK • OXFORD
PARIS • SAN DIEGO • SAN FRANCISCO • SINGAPORE • SYDNEY • TOKYO

Elsevier Butterworth-Heinemann
Linacre House, Jordan Hill, Oxford OX2 8DP
30 Corporate Drive, Burlington, MA 01803

First published 1994
Second edition 2000
Reprinted 2001, 2002, 2003 (twice), 2004 (twice)

Permissions may be sought directly from Elsevier's Science & Technology Rights
Department in Oxford, UK: phone: (+44) 1865 843830, fax: (+44) 1865 853333,
e-mail: permissions@elsevier.co.uk. You may also complete your request on-line via
the Elsevier homepage (http://www.elsevier.com), by selecting 'Customer Support'
and then 'Obtaining Permissions'

British Library Cataloguing in Publication Data
A catalogue record for this book is available from the British Library

ISBN 0 340 75899 6

For information on all Elsevier Butterworth-Heinemann
publications visit our website at http://books.elsevier.com

Set in 10/12 Frutiger
Composition by Scribe Design, Gillingham, Kent
Printed and bound in Malta by Gutenberg Press

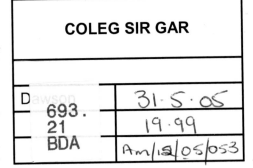

CONTENTS

FOREWORD

Between 1990 and 1993 the Brick Development Association published a workshop manual for the training of bricklayers entitled *Achieving Successful Brickwork*. It was published in sections and is intended to provide detailed support material to assist craft tutors in colleges and other training establishments involved in training bricklayers up to established levels of craft competence so that they can work in the construction industry.

Achieving Successful Brickwork does not define standards for craft skills, but it does describe the procedures that must be followed to achieve good quality brickwork. It also provides a commentary on the technical aspects of design detail, the specification of masonry materials and accessories, and product quality and manufacture – all subjects about which a conscientious craftperson should have some knowledge and appreciation.

Several authors are responsible for the total publication. Individuals with specialist knowledge have been called upon to produce the information on the wide range of topics covered. Guidance regarding bricklaying skills has all been produced by experienced craft tutors. Architects and structural engineers have produced information dealing with the detailed design of modern masonry, and manufacturers have written about brick production methods.

The original manual was greatly appreciated by tutors and many others in the construction industry involved with the building and supervision of brick masonry. Its full colour production, looseleaf/ring-binder format is useful in the training workshop but is unwieldy for convenient reference and relatively expensive for individual students to purchase. There was widespread support for the production of a softback bound, low-priced edition and this present version has been published to meet this demand. It contains the text, diagrams and illustrations in the sequence of the original publication. Although there is a logical order to the material, and the sections do build to a coordinated and consistent whole, it is not envisaged that the book will be read as a continuous narrative but as a reference for particular topics as required. Each section is therefore relatively self-contained and there are some minor instances of repetition to save the inconvenience of frequent cross-references.

FOREWORD TO THE SECOND EDITION

In the five years since the first publication of this book it has won widespread admiration as both a training aid for craft teaching and a technical guidance reference for building professionals, particularly those involved with the detailed design of building and the supervision of construction. The book is now regarded as a standard work of reference in this field.

This new edition is the result of a complete review of the content of the original version. Minor amendments have been made to bring the original text up to date and the opportunity to add new material has been taken. Sections dealing with domestic fireplace and chimney construction are now included.

In sponsoring the production of this new edition, the Brick Development Association reaffirms its commitment to supporting the training of skilled bricklayers and promoting the appreciation and implementation of good practice in the design and construction of brickwork masonry.

TRAINING AS A BRICKLAYER

During the period of the publication of *Achieving Successful Brickwork* the national approach to the formal recognition of craft competence and associated practical and theoretical examination procedures underwent a major change. Formerly, courses of training involved a fixed period of study and practice based on traditional apprenticeship concepts. The new approach, which removes the obligatory fixed time element, assesses skill by ability to demonstrate competence.

The new approach now totally replaces the former one and leads to a National Vocational Qualification (NVQ). NVQs apply to a wide range of industrial and commercial activities and are not exclusive to the construction industry. Eventually the intention is that they will be extended to all vocational pursuits. An NVQ is a measure of competence of an individual's capability to carry out a range of work to performance criteria which had been agreed by Industry. An NVQ is comprised of a number of Units of Competence which can be accumulated over any period of time and in any sequence.

Such qualifications are accessible to everyone. Traditional barriers such as age, duration of training, mode of training, where and how skills had been acquired, are removed. The only constraint remaining is compliance with statutory regulations and legal requirements, e.g. some tasks can only be performed by individuals above a minimum age.

The Construction Industry Training Board (CITB) is the body that has been responsible for defining the range of craft occupations within the construction industry and for establishing definitions and standards of competence for each occupation. It has also reviewed formerly existing qualification procedures and identified appropriate arrangements for assessing work and awarding NVQs.

The CITB and the City and Guilds Institute of London – the body that formerly awarded recognition of craftpersonship ability – are now jointly responsible for assessment and awarding qualifications.

In theory an individual who is able to demonstrate competence in the skills and knowledge defined as necessary for a particular craft can seek assessment and become qualified. However, in practice, most individuals will undertake a formal training programme which involves tuition and practical work as well as ancillary studies and this will be done through a college or other training establishment.

This book does not specifically identify the tasks covered in the various Units of Competence defined and assessed under the NVQ scheme, but all the information and craft guidance is complementary to, and in accordance with, the accepted standards adopted by the awarding bodies and therefore it may be relied upon as up-to-date and authoritative guidance on the construction of modern brick masonry.

THE BRICK DEVELOPMENT ASSOCIATION

The Brick Development Association acknowledges the help and assistance provided by the following persons in the preparation of the written material featured in this publication.

Bob Baldwin PPGB
Consultant

Stuart Bell DipArch, RIBA, MICeram
Technical Director, Marshalls Clay Products Ltd

George Britton ABL
Waltham Forest College

Stephen Brown MMGB
South Cheshire College

Bert Buckley MMGB
The Bournemouth & Poole College

Brian Carling MCIOB, MBIM, Dip.Ed.
New College, Durham

Martin Crosby
Redland Bricks Ltd

Mark Dacey
Pontypridd Technical College

Ray Daniel PPGB
Cumbria College

Graham Foster MMGB, LCG
Stoke on Trent College

Terry Knight AA Dipl, ARIBA
Terry Knight Consultancy

Keith Lamb MMGB
Hull College of Further Education

Mick Lang ABL
Lewisham College, London

Graham Law C.Eng., M.I.Mech.E., M.I.Struct.E. *Technical Director, ANCON Stainless Steel Fixings Ltd*

Mick Pearce JVP, Associate CIOB
Barnsley College, Yorkshire

Graham R. Pellatt MMGB
Highbury College, Portsmouth

David Pope MGB
Langley College, Bucks

Chris Powell MICeram, MIHT, ACIOB
Brick Development Association

Mick Procner
Oaklands College, St Albans

Malcolm Rawson MMGB
Leeds College

Malcolm Thorpe PPGB, MIOB, MBIM, Dip.Ed.
West Nottinghamshire College

Clive H. Wallace MMGB, LICW, LCG
Worcester College

Graham Wright
Leicester Southfields College

Brian Wroe MGB
Wakefield College

Photographs and illustrations used in this manual have been kindly supplied by the following organisations.

Advanced Pro Tools Ltd
Ancon CCL Ltd
ARC Aggregates
British Cement Association
Redland Bricks Ltd
Brick Development Association

Lead Sheet Association
M. Procner
Blakes Building Profiles
Marshalls Clay Products Ltd
Butterley Brick Ltd
Alan Blanc DipArch, FRIBA

Ryarsh Brick Ltd
Terry Knight Consultancy
R. J. Baldwin
D. Pope
Red Bank Manufacturing Co Ltd

Technical Editor Terry Knight.
Design & origination of first edition Barrett Howe Group Ltd, Windsor, Berks.

Additional text and technical co-ordination of soft bound editions
Michael Hammett DipArch, ARIBA, *Senior Architect, Brick Development Association*

BDA MEMBER COMPANIES

Ambion Brick Co Limited
Baggeridge Brick plc
Beacon Hill Brick Company Ltd
Blockleys Brick Limited
Bovingdon Brickworks Ltd
Broadmoor Brickworks Ltd
The Bulmer Brick & Tile Co Ltd
Carlton Main Brickworks Ltd
Charnwood Forest Brick & Tile
  Works Limited
Cheshire Brickmakers Ltd
Chelwood Brick Limited

Chiddingstone Brickworks Ltd
Coleford Brick & Tile Co Ltd
W.H. Collier Limited
Dennis Ruabon Limited
Errol Brick Company Limited
Freshfield Lane Brickworks
  Limited
Hammill Brick Limited
Hanson Bricks Europe
Ibstock Building Products Ltd
Kingscourt Brick
Marshalls Clay Products Ltd

New Forest Brick Ltd
Normanton Brick Co Ltd
Northcot Brick Limited
Ormonde Brick Ltd
Phoenix Brick Company Ltd
Red Bank Manufacturing Co Ltd
Redland Tile & Brick
Sussex Brick Ltd
The Wemyss Brick Co Ltd
The York Handmade Brick Co Ltd

GLOSSARY OF TERMS

*Terms printed in *italics* in the definitions are separately defined within this glossary.

Actual size   the size of an individual brick or block as measured on site. It may vary from the work size within certain allowances for tolerance (see also *co-ordinating size* and *work size*)

Air entrainer   see *plasticiser*

Angle grinder   a powered hand tool with a cutting/grinding disc used for cutting bricks or blocks and also for cutting and *chasing* brickwork or blockwork

Angle support   steel angle fixed to a steel or concrete frame, usually at each floor level, to support brickwork cladding

Angles   *special* shape bricks which form non-right angled corners in walls

Arch   an assembly of bricks which spans an opening in a wall. It is usually curved in form, but may be practically flat

Arris   any straight edge of a brick formed by the junction of its faces

ATR's   'as they rise', a traditional term referring to *stock bricks* that are drawn from the *clamp* or *kiln* and delivered to site unsorted for quality

Autoclave   a pressure vessel used in the manufacture of *calcium silicate* bricks in which they are subjected to super heated steam at high pressure

Axed arch   an *arch* formed of bricks cut to appropriate wedge shape by the bricklayer (see also *gauged arch*)

Band course   a single *course* of bricks forming a decorative contrast of brick colour, *bonding* or shape

Bat   a part brick, e.g. half-brick, three-quarter brick, used in *bonding* brickwork at corners and ends of walls

Batching   the accurate proportioning of *mortar* materials to produce a specified mortar mix

Bed   the horizontal layer of *mortar* on which a brick is laid

Bed face the face(s) of a brick usually laid in contact with a mortar *bed*

Bed joint   a horizontal joint in brickwork

Bench saw   a power-driven, circular saw mounted on a bench which has facilities for holding a brick or block

Benching   floor of a *manhole* or *inspection chamber* formed to discharge into the drainage channel

Bevel (in full bevel square)   a tool with an adjustable steel blade for marking and checking angles when setting out brickwork and laying bricks

Bolster   a broad-bladed chisel of hardened steel used for cutting brick

Bond (1)   the arrangement of bricks in brickwork, usually interlocking, to distribute load

Bond (2)   the resistance to displacement of individual bricks in a wall provided by the adhesive function of *mortar*

Bonding bricks   part bricks, e.g. half- or three-quarter bricks, or specially shaped units to facilitate *bonding* of brickwork at features, corners and ends of walls (see also *bat*)

Boulder clay   a type of clay formed by glacial action. It contains mixed sizes of particles from fine clays to boulders

Brick   see *calcium silicate, clay, common, concrete, engineering, extruded wire-cut, facing, fletton, perforated, pressed, semi-dry pressed, soft-mud, stock*

Brick gauge   a wooden tool to assist the accurate marking for cutting bricks to half- and three-quarters

Brickearth   silty clay or loam in a shallow deposit. Traditionally used for making clay bricks

British Standards   national standards defining the sizes and properties of materials and their proper use in building

Broken bond   the use of part bricks to make good a *bonding* pattern where dimensions do not allow regularised bond patterns of full bricks

Bullnose   *special* shaped brick with a curved surface joining two adjacent faces

Bull's eye   a circular opening in brickwork formed with a complete ring of *voussoirs*

Calcium silicate brick   a brick made from lime and sand (sandlime) and possibly with the addition of crushed flint (flintlime), *autoclaved* in steam under high pressure

Cant   *special* shaped brick with a splayed surface joining two adjacent faces

Capping   construction or component at the top of a wall or *parapet* not providing a weathered overhang (see also *coping*)

Cavity batten   a timber batten, with lifting wires, sized to temporarily lie in the void of a *cavity wall* to catch *mortar* droppings and assist their removal

Cavity tray   see *DPC tray*

Cavity wall   wall of two *leaves* effectively tied together with *wall ties* with a space between them, usually at least 50mm wide

Cellular block concrete block with large voids that do not pass right through the unit

Cement see *Portland cement* and *masonry cement*

Centring temporary former to support underside of *arch* during construction

C&GIL City and Guilds Institute of London. A national training authority

Chases recesses cut in walls to accommodate service cables or pipes

cill see *sill*

CITB Construction Industry Training Board. A national training authority

Clamp a large stack of moulded, dried clay bricks and crushed fuel which is set alight and burns to fire the bricks

Clay brick a brick made from clay formed in a moist condition, dried and fired in a *kiln* or *clamp* to produce a hard semi-vitreous unit

Closers bricks cut to expose a half *header* in the surface of a wall and used as *bonding bricks*

Club hammer heavy hammer used for striking *bolster* when cutting bricks

Collar joint a continuous vertical joint, parallel to the face of a wall, formed in walls one-brick or more thick when bricks are *bonded* in leaves of stretcher bond

Common brick a brick for general purpose applications where appearance is not of significance

Compressive strength the average value of the crushing strengths of a sample of bricks tested to assess load bearing capability

Concrete a mixture of *sand*, gravel, *cement* and water that sets and hardens

Concrete brick a brick made from crushed rock aggregate bound with *Portland cement*

Co-ordinating size size of a co-ordinating space allocated to a brick or block, including allowance for mortar joints (see also *work size* and *actual size*)

Coping construction, or component, at the top of a wall or *parapet* that is weathered, grooved and overhangs the wall surface below to throw water clear and provide protection against saturation (see also *capping*)

Corbel a feature, or *course*, or courses of brick, projecting from the face of the wall, often forming a support

Corner block a wooden, or plastic, block to provide a temporary fixing at the ends of a wall for a string *line* used to control levelling of bricks or blocks when building

Course a row of bricks laid on a mortar *bed* jointed in mortar, generally horizontally

Course stuff a mixture of *sand* and *lime* to which *cement* and water is added to make *mortar*

Cross joint vertical mortar joint at right angles to the face of the wall (sometimes incorrectly called a *perp*)

Datum a fixed reference point from which levels are set out

Diaper decorative pattern of diagonal intersections or diamond shapes produced by contrasting coloured bricks in a *bond* arrangement

Dogleg *special* shaped angle brick

DPC a layer or strip of impervious material placed in a joint of a wall chimney or similar construction to prevent the passage of water

DPC brick *clay brick* of specified maximum *water absorption* of which two courses may be built at the base of a wall to prevent the upward movement of moisture

DPC tray a wide *DPC* bedded in the outer leaf, stepping in the cavity of a *cavity wall*, and built into the inner leaf. It diverts water in the cavity through *weep holes* in the outer leaf

DPM a layer or sheet of impervious material within or below a floor, or vertically within or on a wall, to prevent the passage of moisture

Durability the ability of materials to withstand the potentially destructive action of freezing conditions and chemical reactions when in a saturated state

Eaves lower edge of a pitched roof, or edge of a flat roof

Efflorescence a white powdery deposit on the face of brickwork due to the drying out of soluble salts washed from the bricks following excessive wetting

Elevation a construction drawing showing the view of a vertical surface of a building or object

Engineering brick a type of clay brick traditionally used for civil engineering work for which characteristics of great strength and density are considered beneficial. They are defined by compliance with minimum *compressive strength* and maximum *water absorption* values (stated in the British Standard for Clay Bricks (BS 3921)

Extrados the upper or outer curve of an arch (opposite *intrados*)

Extruded wire-cut bricks bricks formed by forcing stiff moist clay, under pressure, through a die and cutting the extruded shape into individual bricks with taut wires

Face work brickwork or blockwork built neatly and evenly without applied finish

Facing brick a brick for use in the exposed surface of brickwork where consistent and acceptable appearance is required

Fireclay a type of clay containing a high proportion of silica, principally used for the manufacture of fire bricks because of its resistance to high temperature. Also used for building bricks. Generally produces a buff colour

Flashing waterproof sheet material, usually lead, dressed to prevent entry of rain water at an abutment junction between roof and brickwork

Fletton bricks *semi-dry pressed bricks* made from lower Oxford clay, originally made in Fletton, near Peterborough, and subsequently widely used throughout the UK

Flintlime brick see *calcium silicate brick*

Flue a duct, or pipe, that conveys smoke from a fireplace or a heating appliance to the open air

Footing a widening at the base of brickwork to form a bearing on the supporting sub-soil. Traditionally a footing could be of brickwork but in modern construction it is usually of in-situ *concrete* when it is also referred to as a strip foundation

Foundation a sub-structure to bear on supporting sub-soil. May be piles, ground beams, a raft or footings

Frenchman a hand tool used to neatly cut off excess mortar when forming certain mortar joint finishes

Frog an indentation in one or both bed faces of some types of moulded or *pressed bricks*

Frost damage the destructive action of freezing water and thawing ice in saturated materials

Gable portion of a wall above eaves level that encloses the end of a pitched roof

Gauge boxes boxes of specific volumes to accurately measure the proportions of *cement, lime* and *sand* when preparing *mortar*

Gauge rod batten marked at intervals for vertical setting-out of brick *courses*

Gauged arch an *arch* formed of wedge-shaped bricks jointed with non-tapered mortar joints

Gault a clay associated with chalk deposits of East Anglia. Generally bricks made with gault clay are cream or yellow in colour but they may be light red

Gun template *template* shaped to set out angled alignment of *skewback* or *tumbling in*

Handmade bricks see *soft-mud bricks*

Hatching & grinning irregularity of appearance due to the poor vertical alignment of the faces of bricks in a wall surface

Hawk a small board, with a handle in the centre of the underside, used for holding in one hand a small quantify of *mortar* ready for *pointing* with a trowel

Header the end face of a standard brick

Hod a three-sided box, mounted on a pole handle, used over the shoulder for carrying small quantities of bricks or mortar

Hollow block concrete block in which voids run through from *bed face* to bed face

Increment an increase of dimension based on the length of a standard brick

Inspection chamber chamber constructed on a drain, sewer or pipe line with a removable cover to permit inspection, maintenance, clearance of blockages, etc, all when operating from surface level (see also *manhole*)

Insulation batt rectangular unit of resilient fibrous *insulation material* of uniform thickness used to fully fill the air space in a *cavity wall*

Insulation board rectangular unit of rigid *insulation material* of uniform thickness used to partially fill the air space in a *cavity wall*

Insulation material material primarily intended to reduce the passage of heat through a construction

Intrados the lower or inner curve of an arch (opposite *extrados*)

Invert the lowest point on the internal surface of a channel or trough at any cross section

Isometric a drawing, to *scale*, showing an oblique view of an object from a high viewing point

Joint profile the shape of a mortar joint finish

Jointer a tool used to form a mortar *joint profile*

Jointing forming the finished surface profile of a mortar joint by tooling or raking as the work proceeds, without *pointing*

Key brick the central brick at the crown of an arch

Keyed brick a *common brick*, deeply grooved on the *stretcher* and *header* faces as a key for plaster or render

Kiln a permanent enclosure in which clay bricks are fired. There are several designs, some providing for continuous burning

Lateral load force acting horizontally at right angles to the face of a wall. May be due to wind force, retained earth or from associated structure

Leaf one of two parallel walls that are tied together as a *cavity wall*

Level (1) the horizontality of *courses* of brickwork

Level (2) see *spirit level* and *plumb level*

Lime (hydrated) a fine powdered material, with no appreciable setting and hardening properties, used to improve the workability and water retention of *cement*-based *mortars*

Lime (hydraulic) a fine powdered material which when mixed with water slowly sets and hardens and binds together to form a solid material. Traditionally used as a constituent of *mortar*

Lime putty slaked lime, sieved and mixed with water, possibly with a little fine sand, to form a white *mortar*. Traditionally used for thin joints in *gauged arches*

Lime stain (bleed or bloom) white insoluble calcarious deposits on the face of brickwork derived from *Portland cement* mortars which have been subjected to severe wetting during setting and hardening

Line (1)   a string line used to guide the setting of bricks to *line* and *level*

Line (2)   the straightness of brickwork

Line block   see *corner block*

Lintel   a component of reinforced concrete, steel or timber to support brickwork over an opening

Manhole   an *inspection chamber* that permits the entry of a person

Marl   a type of clay with a natural lime content

Masonry cement   a pre-mixed blend of *Portland cement*, filler material and an *air entrainer* used to mix with *sand* and water to form a complete *mortar*

Mortar   a mixture of *sand, cement* or *lime*, or a combination of both, possibly with the inclusion of an *air entrainer*, that hardens after application and is used for jointing brickwork or as *render*

Movement joint   a continuous horizontal or vertical joint in brickwork filled with compressible material to accommodate movement due to moisture, thermal or structural effects

NVQ   National Vocational Qualification. A formal certification of skill competence

Packs of bricks   bundles of bricks secured by bands or straps to facilitate mechanical handling

Parapet wall   upper part of a wall that bounds a roof, balcony, terrace or bridge

Partition wall   wall within a building to compartmentalise the space within it. It may or may not support floors or roofs

Perforated bricks   *extruded wire-cut* bricks with holes through from *bed face* to bed face

Perpends (perps)   notional vertical lines controlling the verticality of *cross joints* appearing in the face of a wall

Pier   local thickening of a wall to improve its stiffness

Pigments   powdered or liquid materials which may be added to *mortar* mixes in small quantities to modify its colour

Pins   flat bladed nails temporarily pressed into mortar joints to secure bricklayers *line*

Pistol brick   *special* shaped brick with a recess in the lower bed to fit over support angle

Plan   a constructural drawing showing a view of a building or object in a horizontal plane. A floor plan shows the floor area of a building with walls in horizontal *section*

Plasticiser   powdered or liquid admixture added to *mortar* mixes in controlled amounts to improve workability by generating air bubbles. Also known as *air entrainer*

Plinth (1)   visible projection or recess at the base of a wall or *pier*

Plinth (2)   *special* shaped brick chamfered to provide for reduction in thickness between a *plinth* and the rest of a wall

Plugging chisel   a stout chisel with a narrow cutting edge for cutting out hardened *mortar* from a joint between bricks

Plumb   the verticality of brickwork

Plumb level   an instrument to check horizontality or verticality of work, consisting of a long, straight-edged casing fitted with two or more *spirit levels*

Pointing   finishing a mortar joint by raking out part of the *jointing* mortar, filling with additional mortar, and tooling or otherwise working it to form the finished *joint profile*

Polychromatic brickwork decorative patterned work which features bricks of different colours

Portland cement   a fine powdered material which, when mixed with water, sets and binds together to form a hard, solid material. It is used as a component of *mortar* and *concrete*

Post tensioned brickwork *reinforced brickwork* in which the steel is tensioned, usually by means of tightening a nut on a threaded end of rod reinforcement, to artificially compress the brickwork and enhance its resistance to *lateral load*

Pressed bricks   bricks formed by pressing moist clay into shape by hydraulic press

Profile boards   temporary timber boards erected outside the enclosing walls of a structure at corners and used to fix *string* lines when setting-out *foundations* and walls

Profiles   patent metal corner posts which are temporarily set up at the corners and ends of walls to support string *lines* and assist building the brickwork to *line* and *level*

Quoin   the external corner of a wall

Quoin block   concrete block of L shape on plan for maintaining *bond* at corners

Racking back   temporarily finishing each brickwork *course* in its length short of the course below so as to produce a stepped diagonal line to be joined with later work

Radial   *special* shaped brick of curved form for use in brickwork curved on plan

Reference panel   a panel of brickwork built at the commencement of a contract to set standards of appearance and workmanship

Reinforced brickwork   brickwork incorporating steel wire or rods to enhance its resistance to *lateral load*

Render   *mortar* applied to a wall surface as a finish

Repointing   the raking out of old *mortar* and replacing it with new (see also *pointing*)

Retaining wall   a wall that provides lateral support to higher ground at a change of level

Returns   the areas of walling at piers or recesses which are at right angles to the general face of the wall

Reveal   the area of walling at the side of an opening which is at right angles to the general face of the wall

Reverse bond   *bonding* in which asymmetry of pattern is accepted across the width of an opening or at *quoins* of a wall in order to avoid *broken bond* in the work

Rough arch   an *arch* of standard bricks jointed with tapered mortar joints

Sample panel   a panel of brickwork which may be built to compare material and workmanship with those of a *reference panel*

Sand   a fine aggregate which forms the bulk of *mortar*

Sandlime brick   see *calcium silicate brick*

Scaffolding   a temporary framework, usually of tubular steel or aluminium, and timber boards to give access for construction work

Scale   the proportional relationship between a representation of an object on a constructional drawing and its actual size, e.g. 1/10th full size = 1:10 = 1 represents 10

Scutch   a hammer with sharp-edged blade, or comb blade, set at right angles to the line of the handle. Used for trimming a cut brick to shape

Sealant   a stiff fluid material which sets but does not harden. Used to exclude wind-driven rain from *movement joints* and around door and window frames

Section   a constructional drawing showing a view of the cut surface that would be seen if a building or object was cut through, generally vertically

Semi-dry pressed bricks   *clay bricks* formed by pressing semi-dry or damp, ground granular clay into shape by hydraulic press

Shale   a type of clay, often associated with coal measures

Sill   the lower horizontal edge of an opening

Size   see *co-ordinating size*, *work size* and *actual size*

Skewback   brickwork, or *special* shaped block, which provides an inclined surface from which an *arch* springs

Soft-mud bricks   bricks moulded to shape from clay in a moist, mud-like state. Often handmade

Soldier   a brick laid vertically on end with the *stretcher face* showing in the surface of the work

Specials   bricks of special shape or size used for the construction of particular brickwork features

Spirit level   device for checking horizontality or verticality consisting of one or more sealed glass tubes, each containing liquid and an air bubble, mounted in a frame

Spot board   board up to 1 m square on which fresh *mortar* is placed ready for use

Springing   plane at the end of an *arch* which springs from a *skewback*

Squint   *special* brick for the construction of non-right angled corners (see also *angle*)

Stock bricks   *soft-mud bricks*, traditionally handmade, but now often machine moulded

Stop   *special* shaped brick to terminate runs of *plinth*, *bullnose* or *cant bricks*

Stop end   a three-sided box-shaped shoe of *DPC* material sealed to the end of a *DPC tray* to divert the discharge of water

Storey rod   *gauge rod* of storey height with additional marks to indicate features such as *lintel* bearings, *sills*, floor joists, etc

Stretcher   the longer face of a brick showing in the surface of a wall

Strip foundation   See *footing*

Suction rate   the tendency of a brick or block to absorb water from the *mortar* used for its bedding and *jointing*. Dense vitrified bricks have a low suction rate, porous bricks have a higher suction rate

Sulfate attack   the chemical reaction of soluble sulfates from the ground or certain types of bricks with a chemical constituent of *Portland cement* which results in expansion of, and physical damage to, *mortar*

Template full size pattern, usually of rigid sheet material, used as a guide for cutting or setting-out work

Throat (1)   a groove formed in the underside of a *coping*, projecting *sill*, or other projecting feature, parallel with its edge and intended to cause water to drip off at that point and not run back on to the surface of the wall immediately below.

Throat (2)   the narrowed part of a *flue* that is located between the top of the fireplace and the chimney flue

Ties   see *wall ties*

Tingle plate   a metal plate shaped to give intermediate support to a *line* when building long lengths of brickwork

Tolerance   allowable variation between a specified dimension and an actual dimension

Trammel   timber batten, pivoted at one end, used to set out curved work

Trowel   hand tool with a thin flat blade, usually diamond shaped, for applying *mortar*

Tuck pointing   a mortar joint finish sometimes used in the 17th, 18th and 19th century work in which mortar joints are finished flush with the face of the walling, tinted to match the bricks and then scored with a regular pattern of false joints to which thin 'ribbons' of *lime putty* are pressed to create the illusion of finely jointed, accurately set-out brickwork

Tumbling in   a sloping feature formed by bricks laid in *courses* at right angles to the face of the sloped surface

Unit of Competence   formal recognition of competence in a specific task. Several Units build towards an *NVQ*

Verge   sloping edge of a pitched roof

Voussior   a wedge-shaped brick or stone used in a *gauged arch*

Wall joint   vertical mortar joint between bricks within a wall and parallel to its face

Wall ties   a component, made of metal or plastic, either built into the two leaves of a *cavity wall* to link them, or used as a restraint fixing to tie back cladding to a backing

Water absorption   a measure of the density of a brick by calculating the percentage increase in the weight of a saturated brick compared with its dry weight

Weep hole   hole through brickwork, usually an unmortared *cross joint*, through which water can drain to its outer face

Work size   the size of a brick or block specified for its manufacture. It is derived from the *co-ordinating size* less the allowance for mortar joints (see also *co-ordinating size* and *actual size*)

1 PREPARATION AND PROTECTION

This section deals with aspects of bricklaying on site that are often left to chance, or left to others.

Because they do not directly involve the placing of bricks and mortar together in the construction of the actual brickwork required, these matters are sometimes regarded as of little importance – 'optional extras' to be dealt with by someone other than the bricklayer. In reality poorly prepared and presented material will hamper the achievement of good quality work and it is in the best interest of the craftsperson to see that materials are correctly stored and handled before they are used. Similarly, it is in the craftsperson's interest to ensure that when the work is complete it will be respected by other trades and protected from weather or damage while the rest of the work is completed. Although these aspects of work may be undertaken by someone in the building team other than the bricklayer, it is nonetheless important that it shall be done.

The building of reference panels and sample panels is also dealt with in this section. These again are not unnecessary 'extras' but give the chance for everyone involved with the work to agree the standard of quality that can be expected with the particular bricks, mortar, and design details that the building project demands. Time and effort spent in this exploratory work is very worthwhile as it will avoid delays caused by later dissatisfaction, disagreement and possible demolition of unacceptable work.

1.1 REFERENCE AND SAMPLE PANELS

Bricklayers may be asked to build both reference and sample panels on site at various times. Bricklayers who understand why such panels are built as well as how to build them can contribute much to achieving good quality brickwork and avoiding costly delays.

DEFINITIONS

A **reference panel** will be built before the facework begins in order to determine design features or to establish standards of workmanship or the visual acceptability of bricks, or all three.

Sample panels are built from subsequent deliveries of bricks for comparison with those in the original reference panel.

WHY REFERENCE PANELS ARE REQUIRED

Their use can save time and money by helping to avoid or resolve disputes which may arise over the quality of the bricks or brickwork.

They may be required for a number of distinct reasons:

1. For the architect to choose a mortar joint colour and profile to suit the specified facing bricks.
2. To establish and provide, for the duration of the contract, a reference to the standard of brickwork which the contractor can produce regularly and which will be acceptable to the architect.
3. Similarly, to provide a reference for an acceptable level of minor or visible surface blemishes such as small surface cracks, chips, small pebbles and expansive particles of lime in the bricks when they are delivered to site (fig 1.1).

Figure 1.1. A reference panel to establish an acceptable level of minor blemishes and proposed mortar colour.

It is a matter which cannot be judged by examining individual bricks. Further reference is made in both the British Standard for clay bricks[1] and that for calcium silicate bricks[2].

4. Samples of special shaped bricks may be incorporated in the reference panel to enable the designer to consider any slight colour variation between the special and normal standard bricks *(fig 1.2)*.

5. Reference panels may also include special features such as soldier courses or narrow piers so that any problem related to brick tolerances and workmanship can be resolved.

Figure 1.2. *Special shapes in a reference panel.*

Reference panels provide useful continuity in the event of changes of personnel, i.e. bricklayers, architects or site supervisors.

LOCATION OF REFERENCE PANELS

The panel should be built where it will remain throughout the contract, readily accessible for viewing in good natural daylight from a distance of 3 m. It should also remain free from damage by vehicles, plant, mud or dirt.

When deciding on a location for a reference panel allow space for subsequent sample panels which should be orientated the same way as the reference panel so that they can be viewed together in similar lighting conditions as well as from the same distance of 3 m.

SELECTING BRICKS TO BUILD A PANEL

The British Standard for clay bricks recommends the adoption of one of two methods:[3]

1. 'supplied by the manufacturer or supplier so that they are reasonably representative of the average quality of the whole order to be delivered' (this may often be the simplest and most appropriate method) or

2. 'randomly sampled in accordance with clause 9' (of BS 3921). This clause gives precise statistical methods of sampling with which many site personnel may be unfamiliar, in which case it may be advisable to avoid duplication of effort by conducting the sampling in conjunction with the manufacturer or supplier.

The methods are designed to select a representative sample of the bricks delivered. Bricklayers should not attempt to select the best bricks and should discard only those which they would discard in practice during the contract.

BUILDING THE PANEL

Separate panels may be built to satisfy the five requirements listed above or one panel may be required to meet more than one of them.

Depending on the requirements, panels should be built:

- on a firm concrete base and stabilised to prevent its being knocked over. In practice this may mean a 215 × 215 pier at each end of a half-brick thick panel which if large enough to display 100 bricks will not be stable.
- to expose not less than 100 bricks.
- to a standard which can be maintained throughout the contract. No attempt should be made to build an 'exhibition panel'.
- discarding only those bricks which would normally be rejected by the bricklayer during subsequent contruction of the contract brickwork.
- to the specified brick bond.
- using the mortar and joint profile specified.
- to a vertical gauge of 4 courses to 300 mm, unless otherwise specified, and plumb, level and aligned.
- with protection to prevent the top of the wall becoming saturated and stained.
- incorporating the specified DPC 150 mm above the slab level to prevent rising damp staining the brickwork and to provide a demonstration of an agreed method for future reference. Flexible DPCs should be bedded on fresh mortar and not laid dry. *(see Section 4.3 'Damp-proof courses')*

Although the requirements dealt with above and elsewhere in this section are concerned with appearance, panels may be required to explore or demonstrate constructional matters such as the position of wall ties, the fixing of insulation, lintels or DPC trays.

SAMPLE PANELS

Sample panels for comparison with the reference panel may be required to be built if there is a dispute about the quality of the bricks delivered to site.

The sample panels should be built in the same way as the reference panel from bricks randomly sampled in accordance with the British Standard for clay bricks.[4] See the second method described under 'Selecting bricks to build a panel' above.

Sample panels should be located so that they can be readily and effectively compared with the reference panel. It is advisable to build them in the same plane because different lighting conditions can result in different appearances. It should also be possible to view the sample panel from the recommended distance of 3 m. Ideally provision should be made to build sample panels immediately next to the reference panel.

EXAMINING AND ASSESSING REFERENCE AND SAMPLE PANELS

Reference and sample panels should be viewed from about 3 m, as noted in the British Standard[1], and only after some days when the brickwork has dried out, because damp brickwork is usually darker than dry brickwork.

References
(1) BS 3921:1985 'Clay Bncks' – appendix F.2.1.
(2) BS 187:1978 'Calcium silicate' (sandlime and flintlime) bricks, clause 8.
(3) BS 3921:1985 'Clay Bricks' – appendix F.2.2.
(4) Ibid. clause 9.

KEY POINTS

- Build reference panels in 'permanent' position on a firm base.
- Panels must be viewed in good light at 3 m distance.
- Allow for building sample panels close by.
- Use bricks especially supplied by the supplier or select at random.
- Protect panels from saturation.

1.2   PROTECTION OF NEWLY BUILT BRICKWORK

The finished work of skilled, conscientious bricklayers can be ruined for all time if it is not protected. *Protected from what? Why and how?*

RAIN

Bricklayers expect rain or even hail or snow to interrupt their work. They may be only temporarily discomforted but their interrupted work can be permanently disfigured unless protected. The causes are inevitable, the results distressing, but protection is simple, requiring only forethought and preparation.

Even light rain falling on newly built brickwork may saturate the surface of the mortar and cause the very fine particles in the cement, lime and pigments to leach, changing the colour of the mortar and giving rise to patchy brickwork.

The fine particles of free lime in Portland cement and hydrated lime may leach and carbonate on the brickwork as hard, shiny, white crystals. Lime leaching is very difficult to remove unlike most efflorescence which is a soft deposit readily weathered away by rain *(fig 1.3)*.

Prevent saturation

To avoid such damage, protect the brickwork before leaving the site or when rain is imminent. A scaffold board or length of DPC, held in place by a few bricks, will often be sufficient.

Although it is particularly important to prevent rain

entering perforations and frogs, even solid bricks should be protected.

When severe or wind-driven rain is expected it is advisable to take more elaborate precautions.

Figure 1.3. Lime leaching.

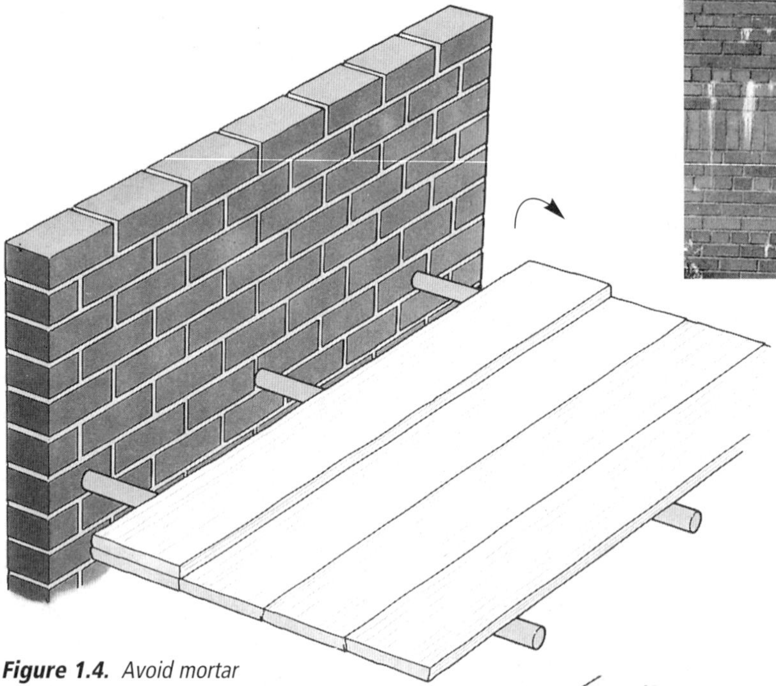

Figure 1.4. Avoid mortar splashing.

Shed rainwater clear of bricks below

Maintain air space

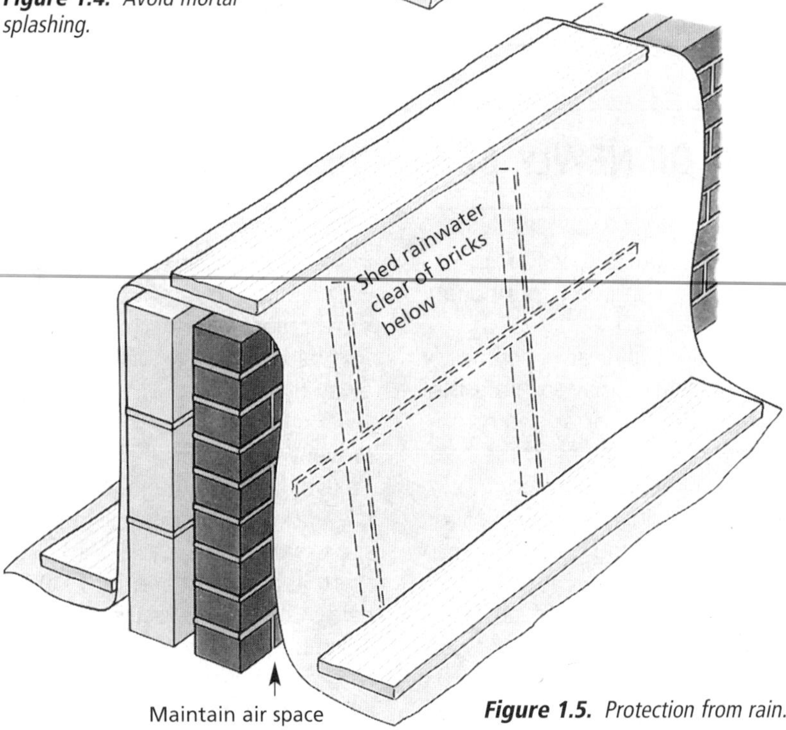

Figure 1.5. Protection from rain.

Polythene sheeting, secured against sudden high winds by scaffold boards or bricks, is effective. Ensure that there is enough to reach down to and protect the lower courses of the new work.

The scaffold boards nearest the brickwork should be turned back if rain is expected otherwise the wall may become stained with bands of mortar splashes which often prove impossible to remove satisfactorily (fig 1.4).

Provide means of maintaining an airspace between the polythene and brickwork as a lack of ventilation may cause condensation which can be as damaging as rain (fig 1.5).

Most competent bricklayers protect their newly finished work but they should also remember to protect bricks stacked on the scaffold, especially if they have a low water absorbency (fig 1.6).

Such bricks can be difficult to lay if saturated. They tend to 'swim' on the mortar bed as they have little initial suction to remove water from the mortar interface.

Figure 1.6. *Protect bricks on scaffolding.*

Also the joints will cure more slowly and when ironed or struck will be of a different colour from those laid with drier bricks thus causing an apparent colour change in the brickwork.

FROST

If mortar is frozen before it has time to set it will remain permanently friable and weak and have poor bond with the bricks. If this happens the brickwork will have to be taken down and rebuilt.

The mortar in newly built brickwork can be protected from freezing by covering with hessian which in turn should be protected by polythene sheeting from becoming wet and useless as insulation. Alternatively, waterproofed insulating sheets will give even better protection *(fig 1.7)*.

SUN AND WIND

In hot, sunny weather, especially with drying winds, the mortar joints may dry before the cement has set and the mortar has bonded adequately with the bricks. This is more likely to happen with bricks of high water absorbency.

This risk can be reduced by carefully draping the brickwork with slightly damp hessian. If the hessian is too wet it may cause staining from the joints *(fig 1.7)*.

PLANT AND PEOPLE

Unprotected brick reveals, sills, arches, thresholds, steps and plinth courses may suffer accidental mechanical damage from building plant or people.

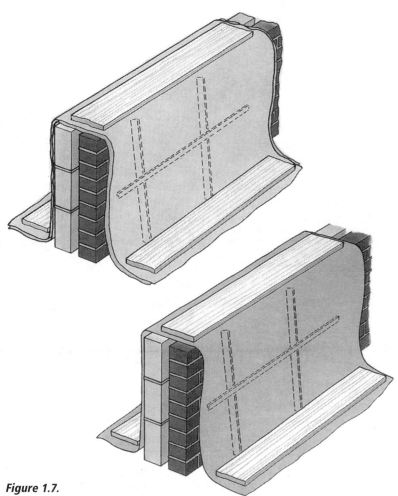

Figure 1.7.

Such damage is not only expensive but often impossible to repair without leaving permanent scars.

Protect such features with strips of plywood or hardboard or in the case of plinths with polythene sheeting *(fig 1.8)*.

WHOSE RESPONSIBILITY?

Site supervisors are responsible for providing protective materials and for giving instructions for their use.

BUT

Ultimately it is the bricklayer's responsibility, as the person on the spot whose work is at risk, to provide timely and effective protection.

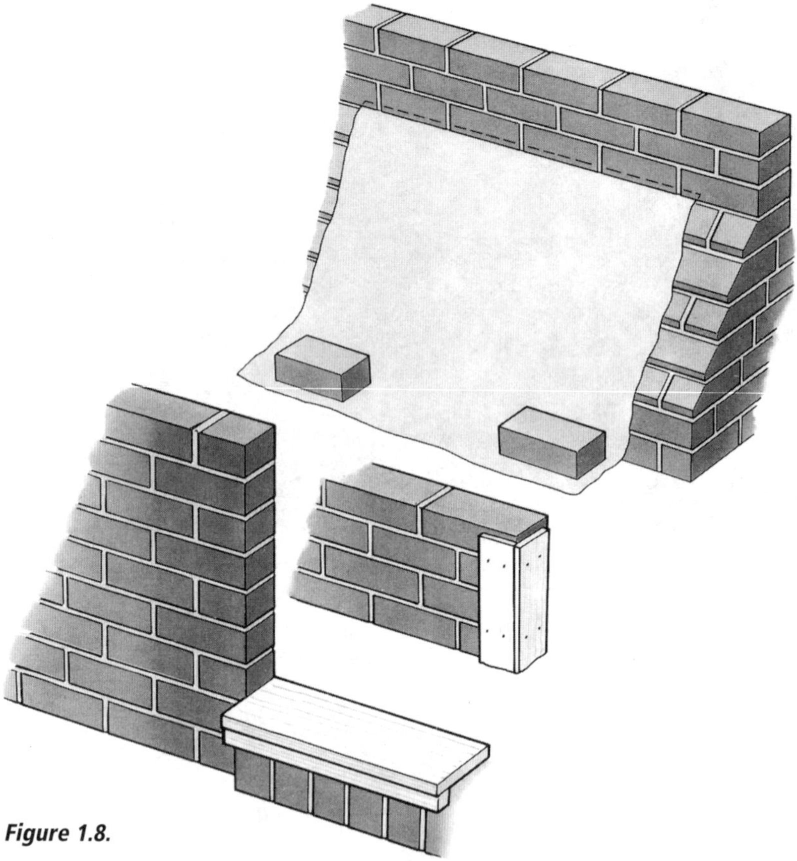

Figure 1.8.

KEY POINTS

- Always protect newly laid brickwork from rain.
- Protect it from frost in winter.
- Protect it from drying out quickly in hot weather.

- Leave an air space between protective sheeting and brickwork.
- Turn back first scaffold board if rain is likely.

- Protect vulnerable brickwork from damage by people and plant.

1.3   HANDLING, STORAGE AND PROTECTION OF MATERIALS

Supervisors and bricklayers are responsible for implementing management policies for safe, careful and efficient handling, storage and protection on site in order to avoid waste of time as well as materials. This section is concerned particularly with avoiding damage to facing bricks.

Because the type of storage and mechanical plant available will vary from site to site it is essential that everyone is aware how handling and storage are to be organised.

STORING AND HANDLING OF BRICKS

There are two basic methods:

- Bricks may be located close to the work place. This reduces the risk of damage and the expense of double handling.
- Alternatively a central compound *(fig 1.9)* provides for better control of pilfering and misuse of all materials inherent in the first method.

Figure 1.9. *Storage compound – bricks, blocks and steel.*

Storage areas must be accessible to delivery vehicles and relevant site plant and the bricks should stand on a firm, level, well drained puddle-free base, not in contact with soil of sulfate-bearing clinker or ashes *(fig 1.10)* nor where they can be mud-splashed by vehicles.

Figure 1.10. *Storage of bricks.*

Figure 1.11. *Special shapes in pack.*

Bricks of special shapes and sizes

'Special' bricks, being particularly valuable, should *always* be stored centrally and easily identifiable for collection as required *(fig 1.11)*. Replacements for many types will not be available 'off the shelf'. *(see 'Storage', p.44, Section 2.9 'Bricks of special shapes and sizes')*

Protection

Generally, leave any polythene wrapping, banding or strapping in place. If the wrapping is removed for inspection or none is provided, replace or provide alternative protection. If the bricks are saturated on delivery provide protection from further rain, but allow air to circulate and dry them before use.

Distribution from a central store

Packs of bricks and blocks should be transported to suitably positioned flat drained areas at work places or for further distribution. If the bricks are for use on upper floors they should be unloaded within the handling radius of tower cranes

Figure 1.12. *Rough terrain fork lift with telescopic mast.*

or placed directly by rough terrain fork lift trucks on a gantry scaffold big enough to take at least three packs of bricks *(fig 1.12)*.

If packs have to be stacked on structural floors they should be placed close to columns away from mid spans whilst allowing space for access and working *(fig 1.13)*.

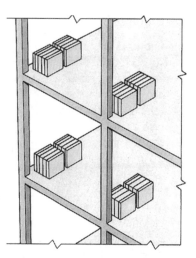

Figure 1.13. *Packs of bricks placed close to columns.*
Note: for simplicity, scaffolding and safety rails are not shown.

A structural engineer should always be consulted before loading out a floor. Typical maximum permissible loading on structural floors and gantry scaffolds is 5–10 kN (½–1 tonne) per m².

NOTE:
- One pack of typical facing bricks is approximately 0.75–1.00 tonne.
- One pack of typical engineering bricks is approximately 1.20 tonnes.

(The above is for information and general guidance only.)

Although the floor above will give some protection to the bricks they will need extra protection from water and wet concrete draining through service holes.

Distribute no more than is necessary for immediate use to any point, but heed any advice from manufacturers to blend bricks by supplying bricklayers from at least three packs. *(see Section 3.2 'Blending facing bricks on site')*

Opening packs
- The correct and safe way to remove banding is by cutting the straps with snips or by placing a chisel under the strap as an anvil and striking with another chisel *(fig 1.14)*.
- A common but not approved method uses two fish-tail wall ties *(fig 1.15)*.
- Do not chop the band with a brick hammer or any tool which will damage the bricks.
- Immediately make banding into safe bundles and remove from the work area. Loose banding entangled round the

Figure 1.14. *Cutting brick banding with 'snips'.*

Figure 1.15. *Incorrect cutting of bands.*

feet can cause serious accidents *(fig 1.16)*.
- Polythene wrapping, removed carefully and put on one side, can provide protection elsewhere.

Loading-out for the bricklayer
- Lay out alternate stacks of bricks and spot boards along the face sides of walls and approximately 600 mm from them.
- Make a level base for each stack, possibly using rejected materials.
- Stack bricks with frog or perforation uppermost, for easy handling by the bricklayer.
- At the end of work protect each stack from rain.
- Supply stacks by drawing from as many packs as possible, but at least three.

Figure 1.16a. *Brick strapping – a 'hazard'.*

Figure 1.16b. *Brick strapping made safe.*

Remove vertical slices of bricks rather than horizontal layers, in order to blend them well *(fig 1.17)*.

Moving to the next working place
- On completion of a lift move all materials to the next work area ready for the next lift. Bricks left on the scaffold might be tipped off and wasted.

HANDLING, STORING AND PROTECTING OTHER MATERIALS

Cement and lime in bags
- Unload without damaging the bags, stack so that consignments can be used in order of delivery and protect

Figure 1.17.

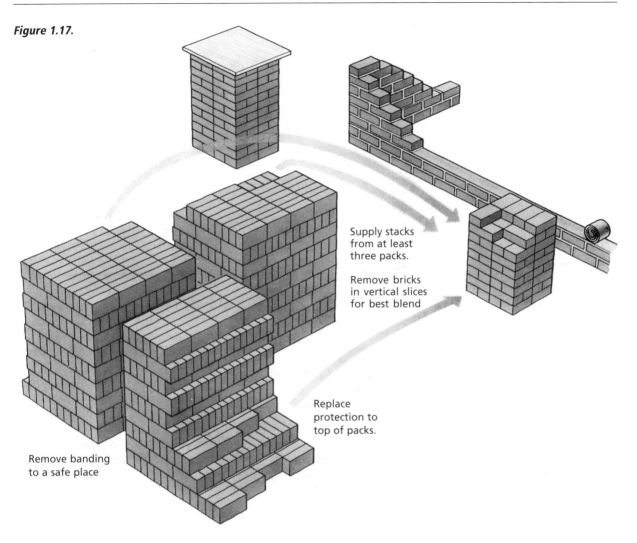

Supply stacks from at least three packs.

Remove bricks in vertical slices for best blend

Replace protection to top of packs.

Remove banding to a safe place

from rain, frost and soil and damp walls. Both lime and cement deteriorate quickly when damp, producing mortar with inferior strength, adhesion and durability. *(see Section 4.1 'Mortars')*

Sand and premixed lime:sand (coarse stuff) for mortars
- Store on a hard, clean, drained base, separating different types of sand and protect from rain especially if frost is imminent. Contaminated sand and coarse stuff may produce mortars with inferior strength, adhesion, durability and appearance.
- Coarse stuff must also be protected from rain and wind, which may erode the fine particles of cement, lime and pigments causing a marked change in mortar colour. *(see Section 4.1 'Mortars')*

Flexible damp-proof courses
- Store rolls on end, no more than three packs high to avoid distortion.

- Protect bitumen and other thermoplastic materials from direct heat.
- In cold weather store sufficient overnight in a warm place for use the following day as some DPCs are difficult to roll out when cold. *(see Section 4.3 'Damp-proof courses')*

Ancillary components
- Lintels, wall ties, thermal insulation batts and boards, DPC adhesives, movement joint fillers and sealants are

examples of materials which should be handled, stored and protected with care in order to avoid damage, loss, distortion and deterioration.

- Read and carefully follow the manufacturers' instructions.

- Additional wooden pallets may be necessary for transferring bags of cement, lintels, DPCs or other materials from opened packs in the central compound.

Further reading
- BS 8000:Part 3:1989 'Workmanship on building sites – Code of practice for masonry'
- Brickwork – good site practice' Knight. The Brick Development Association 1991.
- Sections 2.9, 3.2, 4.1, 4.3, 4.4 of this publication.

KEY POINTS

- ■ Central storage permits better control
- ■ Stand bricks on flat, well-drained surfaces.
- ■ Protect bricks from saturation and contamination.

- ■ Take special care of special shaped bricks.
- ■ Locate packs stacked on concrete floors close to supporting columns but always with the advice of a structural or civil engineer.

- ■ Clear away potentially dangerous cut strapping and banding.
- ■ Protect all mortar materials
- ■ Follow manufacturers' instructions for handling and protection of all materials.

1.4   ESTIMATING QUANTITIES OF BRICKS AND MORTAR

The quantities of bricks and mortar in the following tables have been arrived at by calculation assuming that standard bricks with a work size of 215 × 102.5 × 65 mm are used and that the mortar joints are solidly filled and nominally 10 mm wide.

For the mortar, five figures are given for each wall thickness, depending on the form of the bricks being used and how they are laid, i.e.:

1. **Solid bricks**
2. **Perforated wire-cut bricks**
   (It is difficult to estimate how much mortar enters the perforations as this will vary with the pattern and size of the holes. A 5% increase over the figure for solid bricks is assumed)
3. **Bricks with a shallow frog**
   (In which the frog is about

Quantity of bricks and mortar per square metre of wall surface

Wall thickness (mm)	Number of bricks	Mortar (cubic metre)				
		Solid	Perforated wire cut	Shallow frog	Deep frog (frog up)	Deep frog (frog down)
102.5	59.26	0.018	0.019	0.022	0.030	0.023
215	118.52	0.045	0.047	0.054	0.068	0.055
327.5	177.78	0.078	0.082	0.086	0.107	0.088
440	237.04	0.101	0.106	0.118	0.146	0.120

Quantity of mortar per 1000 bricks

Wall thickness (mm)	Mortar (cubic metre)				
	Solid	Perforated wire cut	Shallow frog	Deep frog (frog up)	Deep frog (frog down)
102.5	0.30	0.32	0.37	0.50	0.39
215	0.38	0.40	0.46	0.58	0.47
327.5	0.41	0.43	0.48	0.60	0.49
440	0.42	0.44	0.50	0.62	0.51

5% of the gross volume of the brick)

4. **Bricks with a deep frog, laid frog up** (In which the frog is up to 20% of the gross volume of the brick, e.g. a pressed Fletton brick)

5. **Bricks with a deep frog, laid frog down**

ALLOWANCE FOR HANDLING AND WASTAGE

The quantities of bricks and mortar given in the tables are based on calculation. In practice allowance must be made for handling and wastage. An increase of 5% is generally allowed for the quantity of bricks and 10% for mortar.

A worked example

Estimate the bricks and mortar required to build a freestanding wall 215 mm thick, 25 m long and 1.8 m high from the top of the foundation to the underside of the coping units.

Surface area of the brickwork:
$$25 \text{ m} \times 1.8 \text{ m} = 45 \text{ m}^2$$

Number of perforated wire-cut bricks for 215 mm thickness:

45×118.52	$= 5333.4$
5% wastage allowance	$= \underline{266.6}$
Total	$= 5600.0$

Volume of mortar
– calculated on wall dimensions:

45×0.047	$= 2.115$
10% wastage allowance	$= \underline{0.211}$
Total	$= 2.326 \text{ m}^3$

– calculated on number of bricks:

5333×0.4	$= 2.133$
10% wastage allowance	$= \underline{0.213}$
Total	$= 2.346 \text{ m}^3$

SUMMARY 5600 bricks and $2\frac{1}{3} \text{ m}^3$ mortar

THE VOLUME OF CEMENT IN BAGS

Mortars are normally mixed on site by volume, but cement is supplied by weight in bags of 25 kg.

The density of Ordinary Portland Cement can be 1200–1400 kg/m³. It is generally taken as 1400 kg/m³ if more specific information is not available.

Based on a density of 1400 kg/m³ the volume of cement in a 25 kg bag is 0.0175 m³ (17.5 litres) which is equal in volume to a 260 mm cube.

For mixing mortar on site it is convenient if a box is made equal to the volume of a bag of cement, then this can be used as an accurate measure for the other constituents of a mortar mix, e.g. a 25 kg bag of cement + 1 box measure of hydrated lime + 6 box measures of sand makes a 1:1:6 cement:lime:sand mortar.

Boxes can be made with no fixed bottom to save lifting and tipping contents out (fig 6.32).

REMEMBER that because binder materials (cement and any lime) in a mortar mix occupy the space that naturally occurs between the particles of the sand, the volume of the mortar is the same as that of the sand, **not** the volume of sand plus the volumes of the cement and the lime (if any).

2 BRICKLAYING TECHNIQUES

This section deals with basic bricklaying skills common to all brickwork assemblies – setting out and control of the regularity of the work, the use of tools, forming joints, etc.

2.1 SETTING-OUT FACEWORK – Stretcher half-bond

Face brickwork should be set-out at the lowest practicable level, ideally below finished ground level, before bricklaying begins, otherwise ill-considered decisions may have to be made later regarding bonding and cutting, particularly at window and door openings. The result may be a lasting monument of poor workmanship.

Setting-out facework is normally the responsibility of the supervising bricklayer who will, after consulting the architect, determine the detailed bond pattern and the location of any broken or reverse bond.

Setting-out the *brickwork* is different from setting-out the *building* which is done before excavation begins.

Although the basic principles of setting-out apply to all brickwork, of whatever bond, this section deals specifically with stretcher half-bond only.

OBJECTIVES

One of the main purposes of setting-out facework is to create a matching and balanced appearance of brickwork particularly at the reveals on either side of door and window openings and ends of walls.

An understanding of the relationship between openings and the bond pattern, if shared by bricklayers, site supervisors and architects will minimise disappointments and delays.

BRICK DIMENSIONS

Brickwork should be set out using the **co-ordinating size** of the length of a brick, e.g. 225 mm (215 mm **work size** + 10 mm *nominal* joint for bricks to BS 3921 and BS 187) *(fig 2.1).*

The above co-ordinating and work sizes are those of the bricks most widely used in the UK, but the specification may call for the use of bricks of a size other that that specified in BS 3921 or BS 187. In this case, setting-out

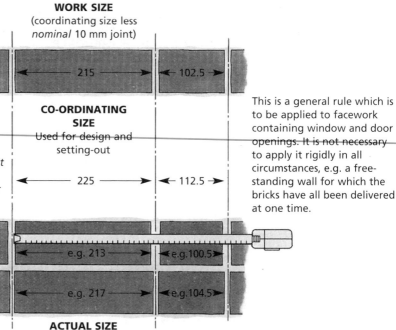

WORK SIZE
(coordinating size less *nominal* 10 mm joint)

215   102.5

CO-ORDINATING SIZE
Used for design and setting-out

225   112.5

e.g. 213   e.g.100.5

e.g. 217   e.g.104.5

ACTUAL SIZE
(as measured)

This is a general rule which is to be applied to facework containing window and door openings. It is not necessary to apply it rigidly in all circumstances, e.g. a free-standing wall for which the bricks have all been delivered at one time.

Figure 2.1.

should be done to a corresponding system of dimensions, e.g. the work size of Metric Modular bricks to BS 6649 'Specification for clay and calcium silicate modular bricks' is 190 × 90 × 65 mm. Again a nominal 10 mm is allowed for mortar joins and their co-ordinating sizes suit lengths in increments of 200 mm and heights in increments of 75 mm.

- If brickwork is set-out using the average **actual size** of the bricks in the first delivery, difficulties may occur if subsequent deliveries differ.

DESIGN

Broken bond and possibly wasteful cutting can be avoided if the overall length of walls and the widths of doors, window openings and brickwork between the openings are all multiples of a brick stretcher. The bonding either side of reveals will also match symmetrically at each course *(fig 2.2)*. (This applies to stretcher half-bond and English bond but not to others such as Flemish and Dutch bond.)

In practice, this ideal situation seldom occurs and a satisfactory solution is dependent on bricklaying skills.

PERPENDS

The bonding pattern should be set out at ground level so that the first few courses establish perpends for the full height of the wall and any problems may be resolved with the architect or site supervisors.

Pencil tick marks made on the brick face should be light and discreet *(fig 2.3)*. Heavy marks over the full height of the brick can spoil the finished brickwork. A test should be made on the type of brick to be used before bricklaying begins as pencil marks are more conspicuous and more difficult to remove on some types of bricks than others.

The verticality of perpends in facing brickwork is as visually important as the horizontality of courses. In practice satisfactory verticality is achieved by plumbing about every fourth or fifth perpend.

REVEALS

The positions of all window openings and 'reveal' bricks should be identified when setting out the first few courses *(fig 2.4)*. This ensures that perpends can continue unbroken for the full height of the wall.

Broken bond

Reveal bricks provide fixed points between which the bonding is set out *(fig 2.5)*. The usually short lengths of brickwork between windows offer little scope for 'adjusting' the widths of cross joints in order to avoid broken bond.

Broken bond can sometimes be avoided by 'tightening' or 'opening' the joints. In doing so bricklayers should work to the

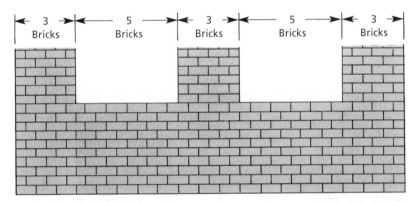

Figure 2.2. *'Ideal' dimensions – whole numbers of bricks and symmetrical reveals.*

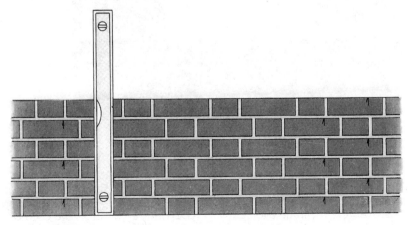

Figure 2.3. *Plumbing perpends.*

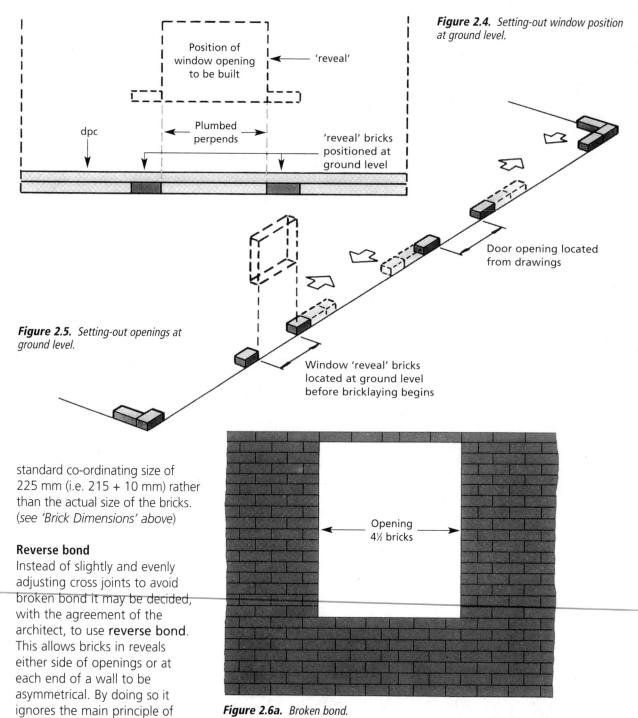

Figure 2.4. *Setting-out window position at ground level.*

Position of window opening to be built

'reveal'

dpc

Plumbed perpends

'reveal' bricks positioned at ground level

Figure 2.5. *Setting-out openings at ground level.*

Door opening located from drawings

Window 'reveal' bricks located at ground level before bricklaying begins

Opening 4½ bricks

Figure 2.6a. *Broken bond.*

standard co-ordinating size of 225 mm (i.e. 215 + 10 mm) rather than the actual size of the bricks. (*see 'Brick Dimensions' above*)

Reverse bond

Instead of slightly and evenly adjusting cross joints to avoid broken bond it may be decided, with the agreement of the architect, to use **reverse bond**. This allows bricks in reveals either side of openings or at each end of a wall to be asymmetrical. By doing so it ignores the main principle of setting out face work, which is to create a balanced appearance, but sometimes architects prefer reverse bond to broken bond.

Broken bond occurs below a window opening, as the result of

setting-out to achieve symmetry of reveal bricks *(fig 2.6a)*.

Broken bond can be avoided by using **reverse bond** *(fig*

2.6b). It is, however, unlikely to be acceptable if contrasting coloured reveal bricks are used a a decorative feature.

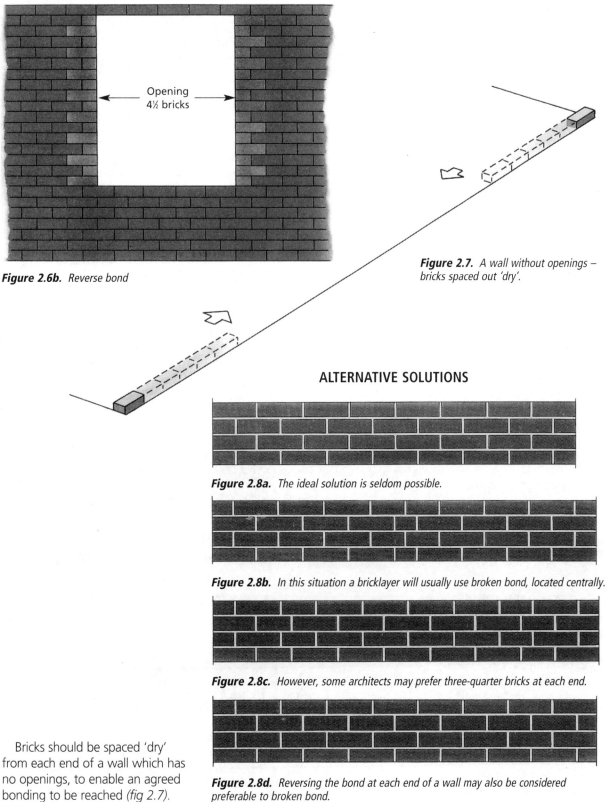

Figure 2.6b. *Reverse bond*

Figure 2.7. *A wall without openings – bricks spaced out 'dry'.*

ALTERNATIVE SOLUTIONS

Figure 2.8a. *The ideal solution is seldom possible.*

Figure 2.8b. *In this situation a bricklayer will usually use broken bond, located centrally.*

Figure 2.8c. *However, some architects may prefer three-quarter bricks at each end.*

Bricks should be spaced 'dry' from each end of a wall which has no openings, to enable an agreed bonding to be reached *(fig 2.7)*.

Figure 2.8d. *Reversing the bond at each end of a wall may also be considered preferable to broken bond.*

POLYCHROMATIC BRICKWORK

Where different coloured bricks are used, especially in band courses, check for differences in average work sizes between them to avoid excessively wide, narrow or badly aligned cross joints. Such problems can usually be avoided by setting-out to 225 mm increments rather than attempting to maintain 10 mm wide cross joints.

KEY POINTS

- Locate the position of openings and the associated reveal bricks above when setting out at ground level and *before* commencing bricklaying.
- Base setting-out at ground level on the 225 mm co-ordinating dimension NOT the actual sizes of the bricks in the first delivery.
- Run out facing bricks 'dry bonded' between the reveals and quoins.
- Plumb the perpends of broken bond from ground level upwards.
- 'Tighten' or 'open' cross joints slightly and evenly in order to avoid broken bond where possible.
- Generally centralise broken bond in walling and below windows unless the use of three-quarter bricks at each end is preferred by the architect.
- Reverse bond may be preferred by some architects in order to avoid broken bond.

2.2   GAUGE AND STOREY RODS

Bricklayers, as part of a construction team, have to co-ordinate or 'work in' with other building components, particularly doors and windows.

One of the most important co-ordinating processes is known as 'keeping the gauge'.

KEEPING TO GAUGE

This refers to working between two given points, A and B, and keeping the bed joints of even thickness *(fig 2.9)*.

Figure 2.9. Gauge and even joint thickness.

STANDARD GAUGE

- The **standard gauge** is **4 courses to 300 mm** (i.e. 4 × 75 mm, the brick co-ordinating size). Standard gauge is normally used when building with the great majority of British standard bricks which are made to BS 392l and BS 187 *(fig 2.10)*.
- These standard bricks are made to a **work size** of 65 mm high (the intended or

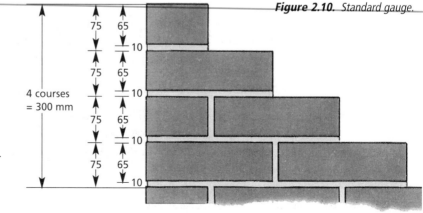

Figure 2.10. Standard gauge.

4 courses = 300 mm

75 65
10
75 65
10
75 65
10
75 65
10

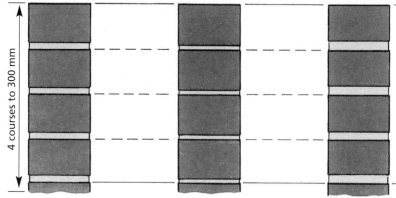

Work size is 65 mm and nominal joint is 10 mm

When average actual size is more than work size, then thinner but regular joints

When average actual size is less than work size, then thicker but regular joints

4 courses to 300 mm

Figure 2.11. *Building to standard gauge – maintaining regular bed joints.*

target size) which is equal to the co-ordinating size of 75 mm less a 10 mm **nominal** joint *(fig 2.10)*.

- The **actual size** will be a little more or less than the work size but the standard gauge must be maintained even though the joints will be a little less or more than the nominal 10 mm. A good bricklayer will ensure that all bed joints are regular in appearance *(fig 2.11)*.

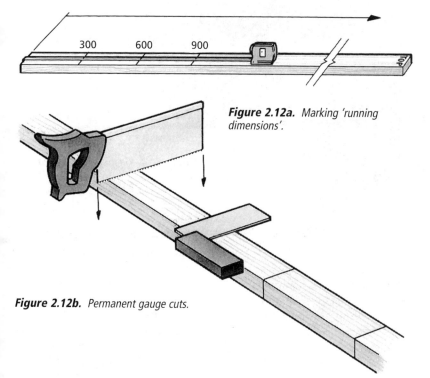

Figure 2.12a. *Marking 'running dimensions'.*

Figure 2.12b. *Permanent gauge cuts.*

- Standard gauge and regular bed joints take precedence over 10 mm joints.

The specification may call for the use of bricks of a size other than the 'standard size' specified in BS 3921 or BS 187. For example thinner bricks, sometimes referred to as Tudor bricks, may be 41 mm high and would be specified to be laid to a gauge of 6 courses to 300 mm, i.e. with a nominal bed joint of 9 mm.

GAUGE RODS

Gauge rods are made and used to maintain gauge and regular bed joints. They are made from a timber lath (typically 50 × 25 mm nominal). The length depends on the height of components such as doors and windows.

The standard gauge is marked on one side and one edge.

Making a gauge rod

Accuracy is essential and is achieved by using running dimensions *(fig 2.12a)*.

- Extend a tape from one end of the lath.
- Make pencil marks every 300 mm (300, 600, 900, etc. as running dimensions).
- Sub-divide these increments into 75 mm increments (75, 150 and 225 mm).
- Use a square to 'square off' the pencil marks at right angles to the sides of the lath *(fig 2.12b)*.
- With a saw make permanent gauge cuts. NOTE: an old woodworking square provides a good guiding edge for the saw.

STOREY RODS

- Storey rods contain other useful information in addition to the standard gauge.
- The heights of walls, window and door heads and sills can be marked on the side opposite to the standard gauge. These are often identified by an inverted V with an abbreviated description. The height above datum is also included as a useful on-site reminder *(fig 2.13)*.
- An alternative method is to mark the bed line saw cut on the same side as the standard gauge and use a colour code, e.g. blue for sills and seatings, and red for lintels and arch springings.

USING GAUGE/STOREY RODS

- Gauging is usually taken from a fixed datum, normally at DPC level *(fig 2.14)*.
- A nail is often simply driven into a joint but a 50 × 25 mm batten screwed into convenient joints gives a more solid base for the gauge rod.
- Check the gauge every course as quoins are raised.
- When checking a quoin for gauge, check the brick for gauge before plumbing and levelling *(fig 2.15a)* because:
  a) if low to gauge, plumbing and levelling would be pointless as the brick must be removed and rebedded.
  b) if the gauge is checked first and found correct then the quoin brick can be levelled, plumbed and aligned without altering the gauge *(fig 2.15b)*.

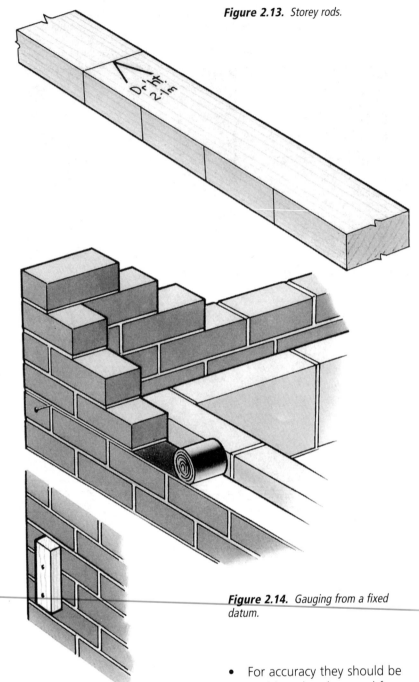

Figure 2.13. *Storey rods.*

Dr. ht. 2·1m

Figure 2.14. *Gauging from a fixed datum.*

'Working in' other components

- A gauge rod, level and straight edge are used to 'work in' and position components such as window and door frames *(fig 2.16a)*.

- For accuracy they should be positioned and gauged from the **top down**.
- Once the frame is positioned and supported the gauge can be marked, on the side towards the bricks, as an extra check when the brickwork is being 'lined in' *(fig 2.16b)*.

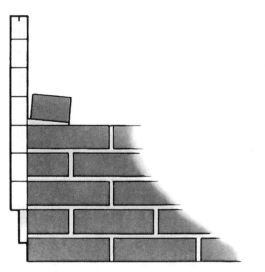

Figure 2. 15a. *Checking gauge of quoin brick.*

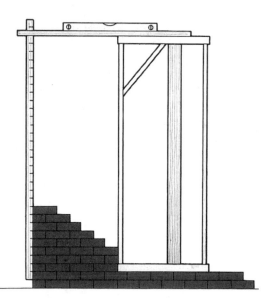

Figure 2.16a. *'Working in' a door frame.*

Figure 2. 15b. *Plumbing quoin brick.*

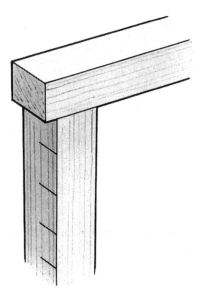

Figure 2.16b. *Gauge marked on back of door frame.*

KEY POINTS

- Use running measurements to mark gauge/storey rod accurately.
- Use a square to mark gauge lines across the lath.
- Set-out the gauge from the top down on components to be built in.

- Store gauge rods flat and dry. They are important pieces of equipment for producing good quality brickwork and should be treated as such.

Existing buildings

If the gauge of new work is required to match that of the existing brickwork, prepare a gauge rod from the existing wall.

Work below ground level

- Drive a peg into the ground to indicate DPC level *(fig 2.17)*.
- With a level, straight edge (if necessary) and a gauge rod, gauge down into the foundation trenches.
- If there is not a whole number of 75 mm courses, any thickened bed joints or split course must be at the bottom of the brickwork.
- The object is to ensure that a bed joint will coincide with DPC level.

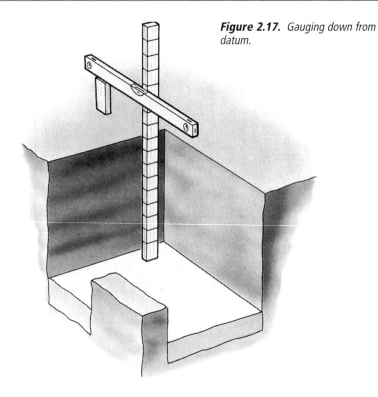

Figure 2.17. *Gauging down from datum.*

2.3   LINE, LEVEL AND PLUMB

Whether building with bricks or blocks the same basic procedures apply.

- The construction of **control points**.
- **Lining-in** between these points *(fig 2.18)*.

These apply both to straight and curved walls, but the procedures detailed in this section are based on straight walls (for curved brickwork see section 5.7).

Quoins must be raised as *control points* before *lining-in* can begin.

BUILDING AN ACCURATE CORNER

When the corners of the brickwork have been marked on the foundation concrete then:

- Lay out materials within easy reach, without obstructing the bricklayer.
- Run out the correct bond, dry, before any bricks are laid.
- Ensure that a datum peg, marked with the DPC level, is within reach of the quoin brick *(fig 2.19)*.
- Lay the quoin brick first. Push it down to gauge and 'level by eye' *(fig 2.20)*. Select reasonably square and regular quoin bricks to make it easier to build an accurate corner.

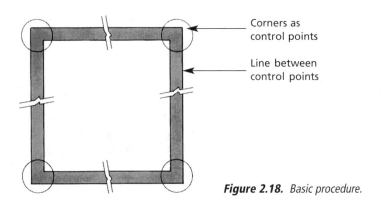

Corners as control points

Line between control points

Figure 2.18. *Basic procedure.*

- Level from the quoin brick, as the first course cannot be plumbed *(fig 2.21)*.
- The number of courses to be raised to complete the corners is the same as the total number of bricks in the first course at the corner *(fig 2.22)*.
- Note that the control point for plumb and gauge is the quoin brick and the more bricks that are laid out from it

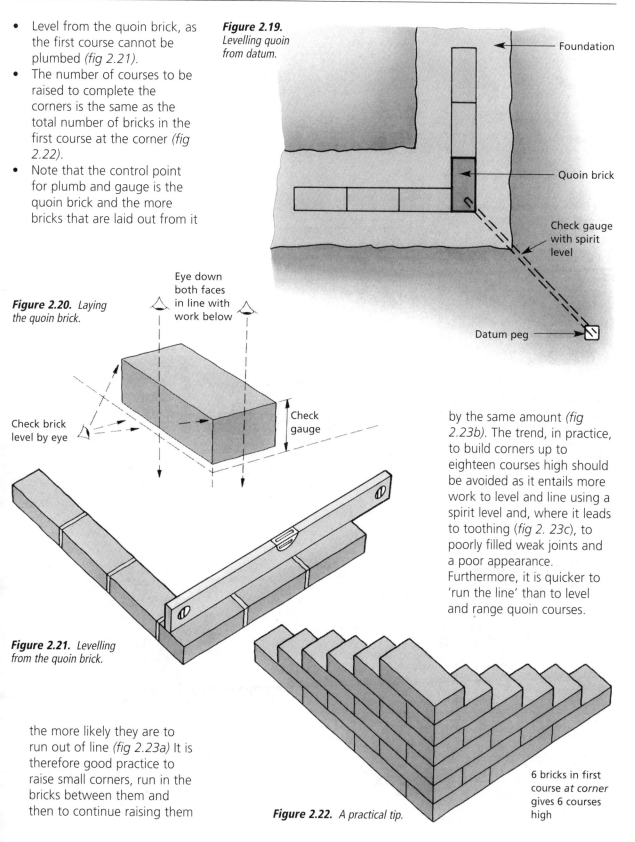

Figure 2.19. *Levelling quoin from datum.*

Foundation

Quoin brick

Check gauge with spirit level

Datum peg

Figure 2.20. *Laying the quoin brick.*

Eye down both faces in line with work below

Check brick level by eye

Check gauge

Figure 2.21. *Levelling from the quoin brick.*

by the same amount *(fig 2.23b)*. The trend, in practice, to build corners up to eighteen courses high should be avoided as it entails more work to level and line using a spirit level and, where it leads to toothing *(fig 2. 23c)*, to poorly filled weak joints and a poor appearance. Furthermore, it is quicker to 'run the line' than to level and range quoin courses.

the more likely they are to run out of line *(fig 2.23a)* It is therefore good practice to raise small corners, run in the bricks between them and then to continue raising them

Figure 2.22. *A practical tip.*

6 bricks in first course *at corner* gives 6 courses high

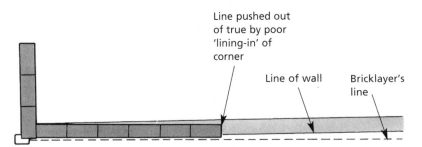

Figure 2.23a. *The danger of long corners.*

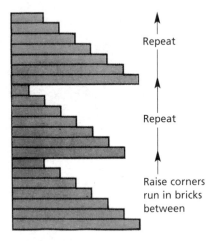

Figure 2.23b. *Building corners – recommended method.*

Figure 2.23c. *Building corners – unsatisfactory method.*

• The second course can be laid, quoin brick first. Keep a check on vertical gauge by levelling out from the quoin brick in both directions at every course. At this stage, plumbing the corner can be only approximate as there are not many bricks to place the level against, so restrain the bottom of the level with the foot and hold the upper part. Always plumb the same part of the corner, otherwise small irregularities in the bricks may cause inaccuracies. Hold the spirit level plumb and gently tap the bricks towards or away from the level, until the full height of the brick, not just one corner, is in contact with the level, and the bubble reads plumb *(fig 2.24)*. Sometimes an irregular brick will have to be knocked out of level to bring its face plumb.

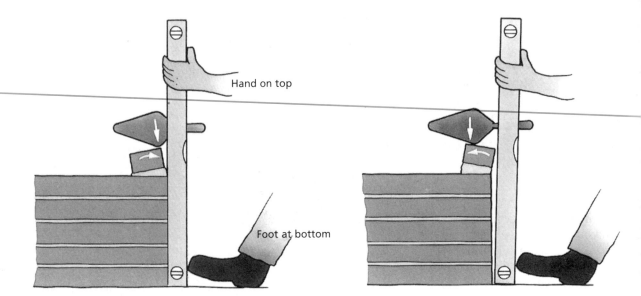

Figure 2.24. *Tapping brick to the level.*   *Tapping brick back.*

Subsequent courses are laid on the corner, racking back as necessary, in the following sequence:

- Bed the quoin brick to gauge and level 'by eye'.
- Check it for gauge and level in both directions.
- Complete each quoin course, levelling from the quoin brick before the mortar bed stiffens.
- Plumb both faces of the quoin.
- Check cross joint thickness and perpends.
- 'Range in' both faces using the spirit level to check face plane alignment (fig 2.25).
- Double-check plumbing on both faces in case they have been disturbed.

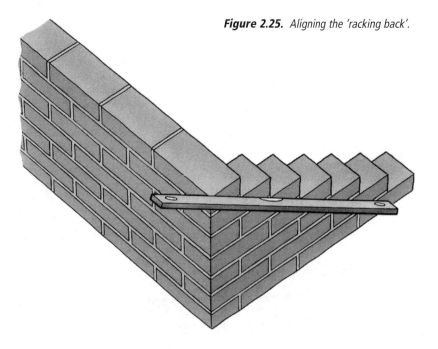

Figure 2.25. *Aligning the 'racking back'.*

The student should consciously 'train his eye' to estimate bed joint thickness, level and plumb accurately to minimise the adjustment needed upon checking with gauge rod and level.

PLUMBING PIERS

When building piers, avoid spreading excessive bedding mortar as the heavy tapping needed will disturb the courses below which will have had little time to stiffen.

With isolated piers it is good practice to get a 'turn of the bubble' batter on both sides. This should prevent the cross joint getting larger and the piers from 'hanging out' towards the top (fig 2.26).

Another method is to plumb once in each direction and to check the opposite face with a tape.

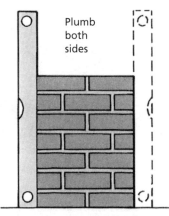

Figure 2.26. *Plumbing piers.*

USING PROFILES

The use of patent profiles eliminates the need for raising quoins and can give greater accuracy because they are placed directly at the control point and the line is fixed to them (fig 2.27).

Note: Patent profiles of various effective designs are produced by different manufacturers. The tool illustrated in diagrams in this book is only intended as a typical example and no inference should be drawn that this particular design is approved or preferred by the authors.

Some other designs have additional featues and/or accessory fittings that extend their usefulness in assisting accurate and well controlled bricklaying.

MAINTAINING LEVEL

The level of the wall is maintained by accurate gauging from the same fixed datum which is usually DPC level. There should be a datum peg at each control point or corner and the bricklayers should work down from them in substructure and up from them in superstructure

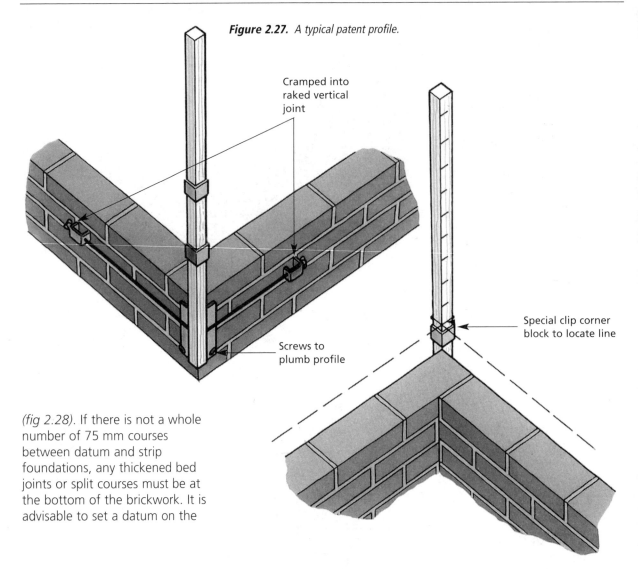

Figure 2.27. *A typical patent profile.*

Cramped into
raked vertical
joint

Special clip corner
block to locate line

Screws to
plumb profile

(fig 2.28). If there is not a whole
number of 75 mm courses
between datum and strip
foundations, any thickened bed
joints or split courses must be at
the bottom of the brickwork. It is
advisable to set a datum on the

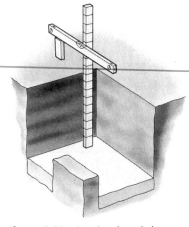

Figure 2.28. *Gauging down below
datum.*

brick wall as soon as possible as
datum pegs often become
dislodged on site.

USING A GAUGE ROD

If the storey height of a building
does not work to gauge, or
existing brickwork has to be
matched, then a 'one-off' gauge
or storey rod should be made up
to suit each unique situation of
non standard gauge. The use of
such a non standard gauge or
storey rod, where this is

necessary, allows the opening or
closing of a gauge to be gradual
and not noticeable. The storey
rod can be marked to show
critical heights such as sill, lintel,
joists and plate levels. More
information on the making and
use of rods is given in section 2.2
'Gauge and storey rods'.

If the corners are built as
shown in *fig 2.29* and the line is
fixed to the same course height
then the wall built between the
corners will be level.

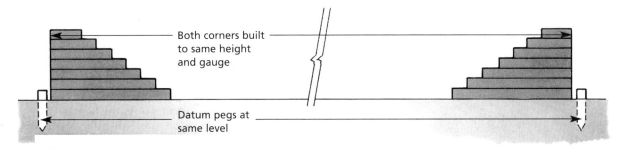

Figure 2.29. *Two corners ready for 'running in'.*

LINING IN PROCEDURES

The line

It is essential to use a good quality line, either of cotton, hemp or nylon. Preferences are a matter for the individual, but for accuracy the line should be light, to prevent sag, and durable so as not to rot if left damp, and without knots.

If lines cannot be spliced, the knot should be wound onto the pin. When the remaining length becomes too short the whole line should be replaced.

The corner block

To avoid 'pin-holes' in a quoin the line can be held to the corner with blocks which are usually made from wood *(fig 2.30)*. The line is pulled through the saw cut and taken once under the line, and through the saw cut again. Hitch the line round the pin to prevent its hanging down too far *(fig 2.31)*.

Place the corner block lengthways to the direction to be run. Keep the line taut to prevent the block from falling. Pull the line through the block as before. The amount of tension necessary to keep the line taut and with no visible sag when sighted, will depend on the length of the wall between corners *(fig 2.32)*.

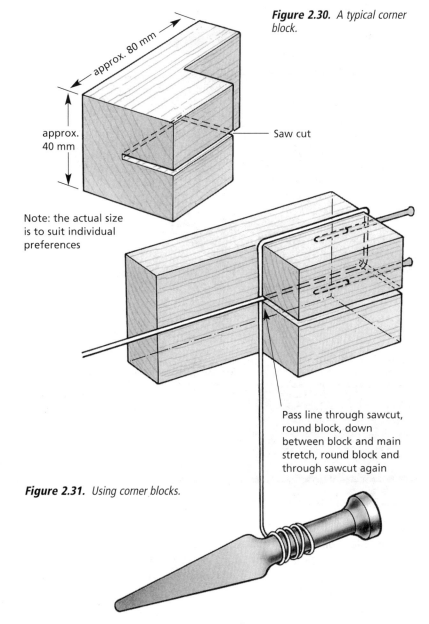

Figure 2.30. *A typical corner block.*

Note: the actual size is to suit individual preferences

Pass line through sawcut, round block, down between block and main stretch, round block and through sawcut again

Figure 2.31. *Using corner blocks.*

Figure 2.32.

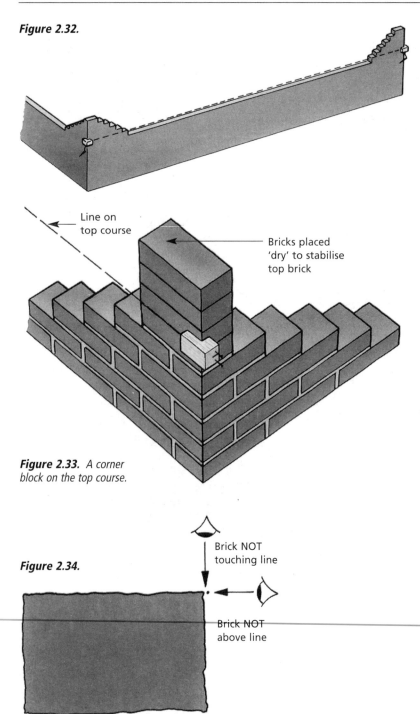

Line on
top course

Bricks placed
'dry' to stabilise
top brick

Figure 2.33. *A corner
block on the top course.*

Figure 2.34.

Brick NOT
touching line

Brick NOT
above line

Checking the general alignment of corners

The bricklayers should check that the quoin courses 'follow the line' each time it is raised.

Running-in to the line

Bricklayers should position themselves so that they can sight both down and across the line *(fig 2.34)*. When eyeing down they should be able 'to just see the light' between the brick and the line. Some bricklayers think of the gap as the thickness of a trowel. **The arris should never touch the line**. The bottom arris should align with the wall face below. When eyeing across the line the top of the brick should not project above the line.

Note: it is usually the 'right hand bricklayer' on a wall who tensions a line.

If the line is stretched over a long length and sags due to its self weight, a tingle plate should be used to support the line as described in section 2.4 'Vertical perpends' and section 5.1 'Copings and Cappings'.

Irregular shaped bricks

Laying some handmade and stock bricks requires experience and skill above that required to lay more regularly shaped bricks. **Surface irregularities must not interfere with the line.**

Internal corners

Here the quoin brick is not the only major plumbing point. The adjacent brick is also important because the line is often fixed 100 mm or more from the internal angle *(fig 2.35)*.

A corner block on the top course

If the mortar is not completely set the top brick may become dislodged when the corner line is raised to the last course, especially on long walls. This can be avoided by weighting the top brick with a few others placed dry *(fig 2.33)*.

A slightly larger corner will need to be raised because the line will need a few bricks as counterweights to prevent dislodgement especially on long walls. On the top course, the line should be taken over the back of the wall and the line held in position by a brick placed dry on the corner (fig 2.36).

Maintaining piers in alignment
Piers in a line should never be built separately. The corner or first piers at each end should be raised ahead of the remainder and a line strung between. The piers should always be a course or so behind the main wall so that a tight line can be pulled through the face of the infilling panels of brickwork (fig 2.37).

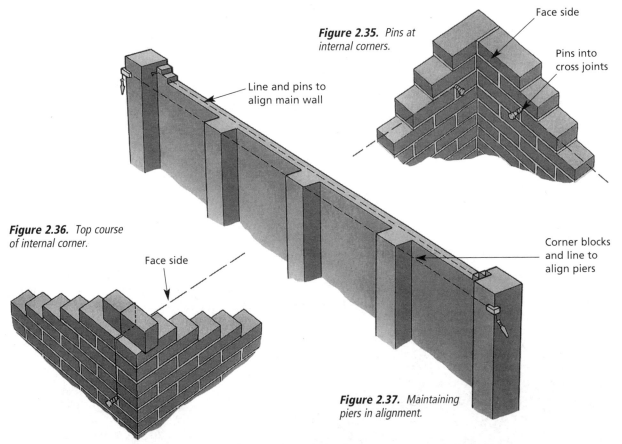

Face side

Figure 2.35. Pins at internal corners.

Pins into cross joints

Line and pins to align main wall

Figure 2.36. Top course of internal corner.

Face side

Corner blocks and line to align piers

Figure 2.37. Maintaining piers in alignment.

KEY POINTS

- Always set out brickwork from profiles or grid lines. Do not just follow the centre of the trench or assume that steel or concrete columns are correctly positioned.
- Run out a wall in dry bricks to locate openings, broken bond and perpend plumbing points before you start laying bricks.
- Always work from datums to keep gauge of brickwork correct all round the building.

- Keep to standard gauge for new work.
- Always make a storey rod in preference to checking with a tape to keep vertical gauge constant
- Raise small quoins throughout a working day.
- Always rack back when raising a quoin. Avoid toothing.
- Keep your foot on the bottom of the spirit level when plumbing a quoin.

- Always keep a tight line.
- Make broken bond as inconspicuous as possible (i.e. two equal cuts over one cut in alternate courses is preferable to a single cut in every course in stretcher bond).
- Plumb perpends every fourth or fifth brick along a course.
- Always plumb the perpends of broken bond.

2.4 VERTICAL PERPENDS

In good quality facework the visible vertical joints in alternate courses should have the general appearance of rising vertically one above the other for the full height of the building without 'wandering'. Although this may appear to be a simple matter, it can be achieved only if supervisors and bricklayers think ahead and exercise care and attention.

DEFINITION

The visible lines of vertical or perpendicular cross joints between bricks are commonly called 'perpends' or 'perps'.

MAINTAINING PERPENDS WHEN RUNNING THE LINE

Once the corners have been built the bricklayers 'run the line', that is they build the walling between the corners. By positioning every fourth or fifth brick exactly above the corresponding bricks in lower courses the 'perps' will remain constant in position.

If the 'perps' are not consciously aligned they may gradually close up on one part of the walling whilst opening on the next part especially if one bricklayer is quicker than the rest. This might result in one bricklayer having to 'crop' or cut bricks to achieve a fit in his part of the wall whilst another may have to increase the thickness of his 'perps' to compensate for the resulting discrepancies.

The method

The difficulties, as described, can easily be overcome with care and forethought.

Firstly, divide the length of the wall into equal sections (normally two) by the use of a tingle, a device for taking up the slack or drop inherent in a line pulled between two distant points. The tingle and its use is illustrated and described in section 5.1 'Copings and Cappings'.

The tingle is placed on a brick bedded near the centre of the run of walling. Plumbing this tingle brick up the wall effectively divides the wall into two sections, causing the bricklayers to adjust only the perpends within their own sections. This will help to prevent the perpends from 'travelling' across the facework (fig 2.38).

Finally, check, at frequent intervals horizontally, that the perpends are vertically above those in the courses below.

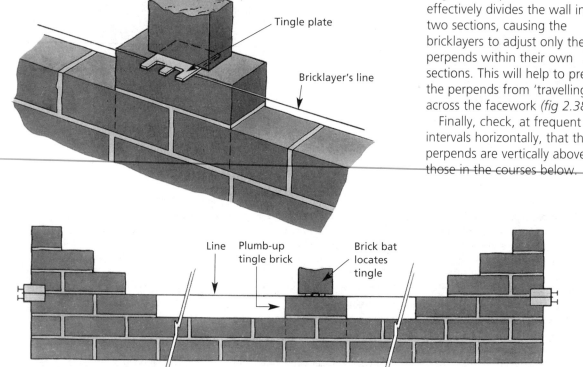

Figure 2.38. Use of a tingle brick to prevent 'travelling' perps.

BELOW GROUND LEVEL

Whenever possible, the actions necessary to maintain vertical perpends should be considered while construction is below ground level and before facework begins. Door and window jambs should be accurately located and the 'reveal' bricks plumbed upwards so that the perpends may be maintained vertically and will not be required to 'travel' across the facework to the correct position for the reveals of a window at a higher level. This matter is more fully covered in section 2.1 'Setting-out facework'.

Once openings and perpends have been identified and located, broken bonds may appear elsewhere. If full length bricks or the intended bonding arrangement cannot be maintained over the length of the wall then some rearrangement will be required. This may result in a new pattern of cut bricks.

PARTICULAR CARE WITH BROKEN BOND

Great care must be taken in aligning the perpends of any pattern of broken bond that has been formed or any other changes in the facework bonding *(fig 2.39a)*. These changes attract the eye and if they contain any misalignment of perpends it will be more noticeable than in the general mass of brickwork such as a flank wall *(fig 2.39b)*.

If the discrepancy in the wall is only slight then it may be better to tighten or open carefully all cross joints rather than to 'crop' bricks. If the perpends are to be aligned correctly then all cut

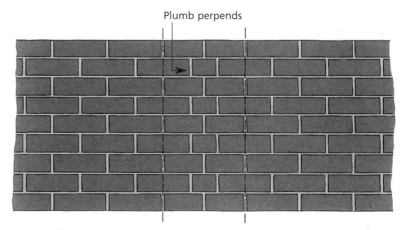

Figure 2.39a. *Care taken to plumb a broken bond.*

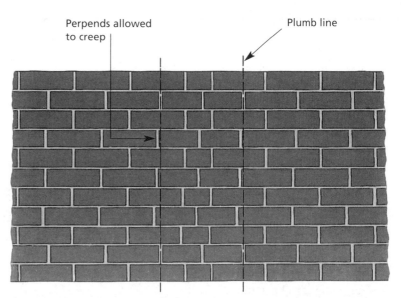

Figure 2.39b. *Poor plumbing at broken bond is visually disturbing.*

bricks should be exactly the same length and the easiest way to do this is to prepare the bricks beforehand using a brick gauge and if available a masonry bench saw.

CHECKING PERPEND ALIGNMENT

Once above DPC, the bricklayer needs to plumb up and mark the arris of every fourth or fifth brick very lightly in pencil. Make a test with the type of brick in use to ensure that the marks can either be removed or will not be visible on the finished brickwork. Particular care may have to be taken with some textured bricks.

Checking perpends can easily be carried out by either of the following methods:

- Place the spirit level on the arris of the lower alternate courses and transfer their positions up the wall *(fig 2.40)*.
- Place a large 'T' square against the arris of the lower alternate courses and transfer their positions up the wall.

Perpends should be checked constantly, starting with the corners. If the perpends on a large corner are 'allowed to wander' it may be impossible to correct the error in the main wall above *(fig 2.41)*. Racking back requires special care because perpend misalignment can very quickly occur. Also perpend misalignment can easily occur when toothing is used *(fig 2.42)*.

Toothing should be avoided as it usually results in poorly filled joints when filling in the toothing. This is likely to increase the amount of water penetrating the outer leaf and be of particular concern to engineers if the brickwork is structural. If toothing is unavoidable, take particular care when filling in.

Perpend alignment, like all bricklaying skills, requires the individual to stand upright occasionally, to step back from his work and look at what he has done and consider its quality before resuming work.

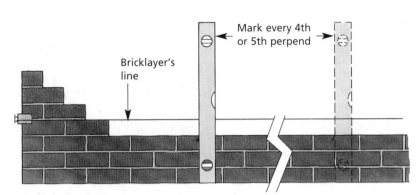

Figure 2.40.

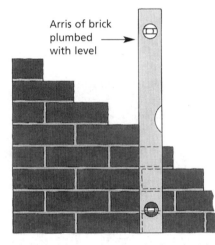

Figure 2.41. *Plumb racking brickwork with care at corners.*

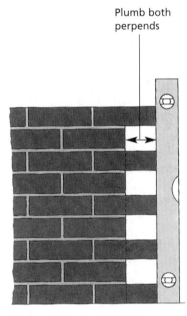

Figure 2.42. *Plumb toothed brickwork with care at corners.*

KEY POINTS

- Consider the actions necessary to maintain vertical perpends before coming out of the ground with facework.
- Where more than one bricklayer is working on a line divide the wall with a tingle.
- Check verticality at frequent intervals.
- Take particular care with perpends when raising corners.
- Take particular care with perpends at broken bond.

2.5   CUTTING BRICKS

One of the best ways of assessing the standard of bricklayers is to examine the accuracy and neatness of their cutting of bricks. High standards depend on skill, care and attention and the use of the correct tools and established techniques.

CUTTING BRICKS – TRADITIONALLY BY HAND

Rough cutting

Trowels may be used for 'rough cutting' but they do not give the necessary accuracy and neatness for facework which needs more appropriate techniques and tools.

Fair cutting

For fair cutting, bricklayers use club hammers, bolsters, comb hammers and brick gauges which they carry as part of their normal tool kit.

Figure 2.43. Brick with fire crack.

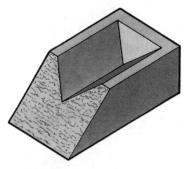

Figure 2.44a. A correctly cut frogged brick.

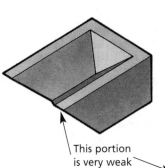

This portion is very weak

Figure 2.44b. An incorrectly cut frogged brick.

Selecting bricks for cutting with hammer and bolster

Select evenly burnt bricks without small fire cracks and other blemishes *(fig 2.43)*. The latter tend to shatter or break in the wrong place under the impact from a hammer and bolster.

Frogged bricks

If frogged bricks are splay cut, the solid bed surfaces should be the longer and the frogged bed the shorter, as in *figure 2.44a* not *figure 2.44b*. The latter will be more likely to break during cutting and laying or fail in use.

Measurement and marking of bricks for cutting

A simple brick gauge *(fig 2.45)* will aid fast, accurate marking when many bricks have to be cut, as for instance when building broken bond in facework. Accurate, clean cutting will help to maintain neat and plumb perpends. The work sizes of brick bats are shown in *figure 2.46*.

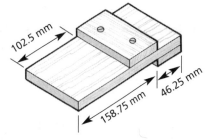

102.5 mm

46.25 mm

158.75 mm

Figure 2.45. A typical brick gauge.

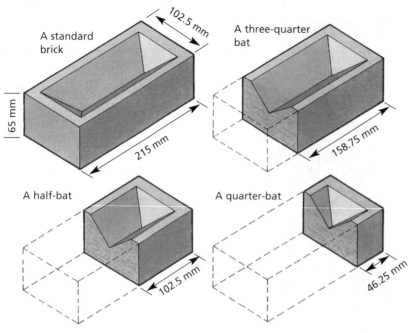

Figure 2.46. *Work sizes of brick bats.*

Bricks should be marked on three sides, the face and opposite side and one bed surface which must be the solid bed of single frogged bricks.

Cutting

Use a small mound or thick layer of sand as a firm cutting pad. The sand, by supporting the brick evenly, minimises the risk of the brick breaking in the wrong place. Alternatives to sand are pads made from DPC off-cuts or from sacking.

- Place the brick on sand or on a pad with the faced side of stretcher uppermost for the first blow *(fig 2.47)*.
- Turn the brick, face down, for the second blow.
- Place the brick flat or frog down for the last blow.
- If the strength of the blows is correctly adjusted, and this only comes with experience,

the third blow will complete the cut.
- If not, repeat the 1st, 2nd and 3rd stages until the brick breaks cleanly.
- To achieve a straight cut on the face side, NEVER start with the brick flat or frog down.
- Trim any rough edges with a scutch or comb hammer.

Figure 2.47. *Cutting a brick.*

Cutting bricks on a scaffold

When bricks have to be cut on a scaffold, make an adequate space, possibly by removing a spot board or stack of bricks. Keep the work areas clear of broken bricks for safe working conditions.

Regulations regarding the hand cutting of bricks

It should be noted that 'Health and Safety' regulations require the wearing of eye protectors when cutting bricks with a hammer and bolster.

CUTTING BRICKS WITH A MASONRY BENCH SAW

The use of masonry bench saws is effective and advisable when large numbers or particular types of bricks have to be cut.

Experience suggests that while 3-hole perforated bricks can be cut satisfactorily by hand it is very difficult to cut multi-hole perforated bricks accurately and consistently without creating a lot of waste. Similarly, the cutting of very high strength, low water absorbency solid bricks will almost certainly require a bench saw *(fig 2.48)*.

Masonry bench saws are particularly useful for cutting bricks at an angle which can be very difficult to do by hand. For further information about bench saws and their location, see section 6.9 'Bricklaying tools and equipment'.

Correct operation

- For safety, keep area around the bench saw free from debris and obstacles.
- Ensure blade is tight on spindle.

***Figure 2.48.** Bench saw.*

- Check trolley moves freely on the rails.
- Ensure an adequate supply of water in the machine.
- Clearly mark bricks to be cut, as described in 'Measurement and marking' in this section.
- Put on ear and eye protectors which are required by law but also wear gloves, waterproof apron and boots.
- Place brick on the sliding table and offer up so that the blade is central over the cutting mark. Secure brick with fixing clamp.
- Turn on power to start machine and check that the blade turns evenly and free from wobble.

- Offer the brick on the sliding table to the rotating blade, apply pressure to the spring-loaded foot pedal to draw the blade down on the brick. It is important to 'let the machine do the work'. A little even pressure, keeping the motor revolutions even, is enough to guide the blade through the brick.
- Keep fingers clear of blade, never look away while using saw.
- If sparks are emitted it usually means insufficient water is reaching the blade.
- When the brick is almost cut through, release pressure so

that chipping does not occur.
- When cutting is complete, switch off, unclamp the brick and leave to dry before use, as, if saturated, the brick faces may become smeared with mortar.
- Change the water before each change of brick. Light, buff bricks, for instance, may become stained if cut after black bricks.

Note: An approved Cautionary Notice (Abrasive Wheels Legislation) must be displayed for easy reading by operatives.

KEY POINTS

- Always use a pile of sand or other pad when fair cutting.
- Prepare a space when cutting on scaffolding.
- Use a gauge for quick, accurate marking.
- Apply bolster to face side first.

- Do not strike too hard with bolster.
- Select technique to suit type of brick.
- Follow all safe working procedures when operating a bench masonry saw.

2.6   KEEPING BRICKWORK CLEAN

Well designed, specified and otherwise skilfully built facing brickwork can become disfigured by mortar stains unless care is taken to work cleanly and protect materials and finished work.

This section is mainly about good trowel technique and control of mortar consistency to minimise such staining. But

it also refers to other techniques, described elsewhere in this book.

TROWEL TECHNIQUE

Cross joints
Apply mortar deftly to the end of bricks so that faces are kept

perfectly clean while filling the cross joints completely to maximise rain resistance *(figs 2.49a & b). (see Section 6.7 'Rain resistance of cavity walls')*

Bed joints
Spread mortar fully across the wall, but let none droop and stain the facework below. Do not deeply furrow *(figs 2.50a & b)*

To avoid smearing the bricks in the course below, gather the surplus 'squeeze' of bed joint mortar produced as each brick is pressed down to the line. Do this with the trowel blade, making a 'cutting' action along the line of the bed joint, horizontally **not** upwards *(figs 2.51a & b)*.

Do not put too much bedding mortar on the wall. *It will increase the risk of wetting the brick face and can lead to excessive tapping which encourages water towards the face. Estimating the correct quantity of bedding mortar and spreading it efficiently is an important training aspect (fig 2.52).*

Point the trowel along the wall when tapping bricks down so that any mortar dropping from the blade falls on the bed *(fig 2.53a)*. If the trowel is held across the wall, mortar can drop on the face *(fig 2.53b)*. Take greater care as you modify your trowel technique to reach higher as the wall rises.

Figure 2.49a. *Apply mortar cleanly to end of bricks.*

Figure 2.49b. *Avoid smearing the face.*

Figure 2.50a. *Acceptable light furrowing.*

Figure 2.50b. *Unacceptable deep furrowing reduces load-bearing capacity.*

MORTARS

Excessively wet mortar, whether site-mixed or delivered in tubs, retarded and ready-to-use *(see Section 4.1 'Mortars')*, is a common cause of dirty brickwork.

In order to press bricks down by hand, bricklayers need smooth workable mortars, but if they contain too much water or plasticising admixtures they will be 'sloppy', causing even the most skilful bricklayers to smear the face.

Bricklayers are responsible for adjusting the consistency and workability of mortars to suit the type of bricks being used. But they must not, without the permission of the architects or engineers, adjust the proportions of cement, lime and sand as this can reduce the strength or durability of mortars.

Figure 2.51a. *Correct cutting action.*

Figure 2.51b. *Incorrect cutting action.*

Figure 2.52. Judge the right quantity of bedding mortar and spread evenly.

Figure 2.53a. Mortar drops on bed of brick.

Figure 2.53b. Mortar may drop down face.

information on bricks see Section 6.10 'Brick manufacture')

Higher absorbency bricks

Providing such bricks are not wet at the time the mortar is spread, they will normally have sufficient suction rate to absorb some water immediately, thereby reducing the risk of mortars running down the facework.

But, if the bricks are extremely dry, as in a hot summer, they may absorb water too quickly. In that case it is better to reduce their suction rate by wetting, rather than making the mortar too wet and 'sloppy'.

Figure 2.54. A typically high absorption brick with a high suction rate when dry.

Sand for mortars

Mortars should be made from well-graded sand containing fine, medium and coarser particles. *(see Section 4.1 'Mortars' fig. 4.3.)* Well-graded sands used in mortars help retain mixing water long enough for the mortar to develop its maximum strength, durability and adhesion. Sand containing only larger particles makes 'hungry' or 'short' mortar, allowing mixing water to 'bleed out' on the spot board and/or down the wall face. Factory-produced mortars from reputable suppliers can be expected to contain suitably

graded sands. Builders' merchants usually refer to such sands as 'building sands'.

Lime in mortars

Lime helps to prevent mortars drying out too rapidly while in use. This is considered to improve cohesion, workability and adhesion and in turn reduce the risk of smearing.

TYPES OF BRICKS

Bricklayers are responsible for adjusting the consistency and workability of mortars to suit the type of bricks. *(For further*

Examples of such bricks are some machine-moulded stock and handmade bricks *(fig 2.54).*

Lower absorbency bricks

Such bricks generally have a low suction rate and absorb very little water from mortars as they are spread. When using such bricks bricklayers should adjust the consistency of mortars to produce a less 'sloppy' mix in order to prevent its running down the face.

Examples of such bricks are extruded wire-cut bricks *(fig 2.55).*

Figure 2.55. *A typical low absorption brick with a low suction rate, especially when wet.*

Textured bricks

Dragwire and other textured brick faces tend to pick up mortar more readily than smooth faces during bricklaying.

To reduce this risk, remove the 'squeeze' of excess mortar in smaller amounts by two or three horizontal passes of the trowel as the brick is squeezed down, rather than a single pass which tends to pile mortar on the face.

OTHER TECHNIQUES

Good trowel technique alone is not enough. Protect bricks from becoming very wet from rain and ground water when in store and loaded-out on the scaffolding. *If medium to low absorbency bricks are very wet it becomes more difficult to avoid smears. (See Section 1.3 'Handling, storage and protection of materials')*

Completed facework should also be protected to prevent rain from washing fine particles of cement, lime and pigments from fresh mortar.

Plinths and other projecting brickwork should be protected from mortar droppings. Scaffold boards next to facework should be cleared of mortar and turned back during rain, and overnight. *(See Sections 1.2 'Protection of newly built brickwork' and 6.6 'Appearance'.)*

CONCLUSION

The necessary cleaning down of facework at the end of a contract can be greatly reduced by good trowel technique and avoiding the use of 'sloppy' mortars.

In addition, vulnerable finished brickwork must be protected from subsequent staining by mortars and materials used by other trades. Mortar must be cleared from scaffold boards and those nearest the brickwork turned back to prevent splashing caused by rain falling on them.

KEY POINTS

- Keep mortar workable but reduce mixing water as much as possible.
- Use well-graded building sand.
- Do not spread bedding mortar too thickly.
- Remove surplus mortar from face with horizontal cutting action of trowel.

- Apply cross-joint mortar carefully to end of bricks.
- Do not tap bricks unnecessarily – press them down to the line.
- Although bricks may have to be wetted in hot dry weather do not lay bricks that are saturated.

2.7   FINISHING MORTAR JOINTS

The type of joint finish and the skill and care with which it is carried out, profoundly affects both the appearance and rain resistance of brickwork. *(see Sections 6.6 'Appearance' and 6.7 'Rain resistance of cavity walls')*

Today, most brickwork is 'jointed', which means that the joints are finished as the bricklaying proceeds. 'Pointing' of mortar joints that were raked out on the day the brickwork was built is not now common. If pointing is specified it is normally carried out after the completion of bricklaying. *(see Section 2.8 'Pointing and repointing')*

Joint finishing is usually left to a convenient moment. 'Joint-up before you break for tea' is the usual reminder from the foreman bricklayer or charge hand. Likewise, bricklaying will generally stop before the end of the day's work to leave time for jointing-up.

This section stresses not only the importance of allowing

sufficient time for finishing joints correctly but the need to do so at the **right** times throughout the working day.

APPEARANCE

'Good pointing can improve poor brickwork but bad pointing can spoil good brickwork,' is a saying well known to experienced bricklayers. Similarly, careful finishing of mortar jointing can minimise the effect of small deficiencies in bricks and bricklaying but careless joint finishing can make them look worse.

Jointing-up is a critical part of building facework and is not something to be 'dashed off' apart from the main operation of bricklaying. It considerably affects the permanent appearance of facework as almost one fifth of the total surface consists of mortar joints.

Probably the most important aspect of jointing-up is to avoid smudging the bricks or blocks. *(see Section 2.6 'Keeping brickwork clean')*

TIMING

Timing is probably the most important aspect of jointing-up, particularly when making a neat flush joint without smudging the facework.

The right time to joint-up is determined by both the suction rate of the bricks and weather conditions at the time the bricks are laid.

At one extreme, bricks of low water absorption that are very wet will have a low suction rate. The bricks will tend to 'float',

and the mortar will dry slowly, especially during wet or cold weather.

At the other extreme, high water absorption bricks that are very dry will have a high suction rate and the mortar will dry out very quickly, possibly before a bond with the bricks has fully developed.

To avoid these extreme conditions all bricks should be protected from saturation, and in hot, dry weather the suction rate of higher absorption bricks should be reduced by docking or lightly spraying so that the surface is left damp rather than wet.

During summer months it is usually necessary to joint-up every two or three courses in a length of walling typically built by one bricklayer. In winter, twice only in a lift of brickwork may be appropriate. The mortar should be 'soft' enough for the jointing tool to leave a smooth surface and to press the mortar into contact with the brick arrises in order to maximise rain resistance.

Trying to finish a mortar joint which is **too** dry, and pressing **too** hard with the jointer can 'blacken' the joint face and leave a crumbly surface *(fig 2.56)*.

Jointing-up too soon spreads the mortar and leaves a rough joint surface *(fig 2.57)* Northamptonshire bricklayers say 'Wait until the joints have "hazelled off" a bit before you joint-up'.

TECHNIQUE

When 'ironing in' to give joints a 'bucket handle' finish for example, each bricklayer must

Figure 2.56. *Jointing-up late can disturb and crumble mortar surface or 'blacken' it by over-rubbing.*

Figure 2.57. *Jointing-up too early can smear mortar and leave a rough surface.*

Figure 2.58. *Correct use of jointer.*

use the **same diameter** jointer for consistency *(fig 2.58)*. The jointing tool must remain in contact with brick arises above

Figure 2.59. *Incorrect use of jointer leaving 'tramlines'.*

Figure 2.60. *Chariot jointer showing spike for raking joints to a consistent depth.*

Figure 2.61. *An improvised but effective depth gauge.*

Figure 2.62. *Forming a weather struck joint.*

and below the bed joints and each side of cross joints, otherwise 'tramlines' will be left *(fig 2.59).*

Cross joints must always be finished first whatever the joint finish. The only exception to this rule is when applying tuck pointing.

If square recessed joint finish is specified, the use of 'chariot' jointers *(fig 2.60)* or other depth gauges *(fig 2.61)* will help each bricklayer to rake out to the same depth. After raking out, the recessed mortar should normally be 'polished-up', not left rough, in order to maximise the rain resistance of the joint. Use the insert in a chariot jointer or a square jointing tool.

If a weather struck finish is required, all cross joints must be inset the same side, usually the left-hand side, by both right and lefthanded bricklayers *(fig 2.62).* Otherwise there will be distinct differences in appearance due to shadow lines.

Note: With weather struck and cut pointing the depth of inset and 'boldness' of mortar projection must be consistent. The thickness of the pointing

trowel is sufficiently bold. *(see Section 2.8 'Pointing and repointing')*

It may appear easy to achieve a high standard of simple 'flush' jointing but experience, care and attention is required if the joints are to be left looking truly flush.

If finished too soon 'wet' mortar will be smeared on the face of the bricks. If left too late, so that the mortar is too 'dry', it will crumble and leave 'misses' and an incomplete and unevenly compacted surface.

Steel tools are used to finish the surface of bucket handle, struck weathered and also give the final polish to square recessed joints. Steel tools are not used to finish flush joints as they tend to leave conspicuous tooling marks. A piece of hardwood, about 200 mm long by 50 mm wide and 10 mm thick with a half-rounded end, rather like a doctor's spatula, is commonly used to flatten the mortar joints. Great care should be taken to ensure that both cross joints and bed joints are left truly flat and not 'dished'.

A very light brushing will remove fine crumbs of mortar and leave a matt surface rather than a polished one as from a steel tool.

The comments on 'brushing', in section 2.8 'Pointing and repointing', applies equally to jointing. Generally, delay brushing until the end of the day in summer and leave until the next day in damp or cold winter weather.

ATTENTION TO DETAIL

Care is needed when finishing joints at external angles *(fig 2.63).* Finishing of joints in internal angles must carefully emphasise the tie-bricks or bonding at these points, finishing alternately to left and right *(fig 2.64),* not with a straight joint *(fig 2.65).*

Take care to continue with the joint finishing under projecting brick-on-edge sills, under copings and the soffits of soldier arches over openings.

Figure 2.63. *Pay attention to external angles.*

Figure 2.64. *Correctly finished internal angle.*

Figure 2.65. *Incorrectly finished internal angle.*

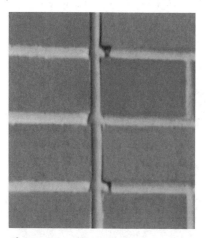

Figure 2.66. *'Mouses' ears' in a vertical movement joint.*

At vertical movement joints in facework leave a vertical, parallel sided 10 mm space for mastic sealant. Do not leave gaps at the ends of bed joints so that the sealant spreads to give unsightly 'mouses' ears' *(fig 2.66).*

MORTAR

Building sand that is predominantly fine grained will produce a closer textured, smoother and denser surface from the action of the jointer than will coarser sands. Lime in bricklaying mortar tends to produce a more compact surface than when air entraining plasticisers are used.

KEY POINTS

- Take care to judge the best time to joint-up.
- All bricklayers should use the same profile jointing tools and technique.
- Take particular care at angles, sills and vertical movement joints.
- Use fine grained sands for fine joint finishing.
- Brush lightly if at all.

2.8 POINTING AND REPOINTING

Immediately after a brick has been laid the surplus mortar is cut flush with the face. The main purpose of any subsequent joint finishing is to improve the rain resistance of the wall by compacting the surface of the mortar and pressing it into close contact with the bricks.

JOINTING

Today most face brickwork is 'jointed' which means that the joints are finished as the work proceeds and should require no further attention at the end of the day.

POINTING

Occasionally architects will specify that the joints be 'pointed' in order to achieve a particular effect.

When new brickwork is to be pointed, all joints are raked 12 to 15 mm deep on the day the wall is built ready to receive a different mortar at a later date *(fig 2.67)*.

The pointing is usually a different colour and may be required to have a particular profile, e.g. flush, struck weathered, half-round tooled or square recessed, all of which can also be formed as a jointing process as the work proceeds *(fig 2.68)*.

In this section, only weather struck and cut pointing will be described, as it is a commonly used profile for pointing and repointing *(fig 2.69)*. Also, it is

Figure 2.67. *Raked out joints.*

not practicable to form the profile as part of the jointing process because the bricklayers' rhythm would be disrupted whilst they stopped to apply, to every joint, extra mortar to form the profile. Forming a weather struck and cut finish is a pointing, not a jointing operation.

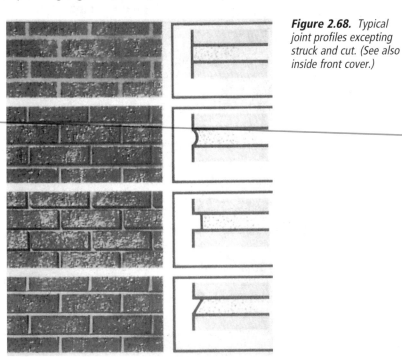

Figure 2.68. *Typical joint profiles excepting struck and cut. (See also inside front cover.)*

TOOLS

Pointing trowels with blades 50, 75, 100 and 150 mm long are used with a hand held hawk. The shortest trowel is known as a 'dotter' *(fig 2.70)*.

THE PROCESS

Pointing is seldom popular with bricklayers, for being a static operation and requiring patience, care and attention it can be a cold job during the winter.

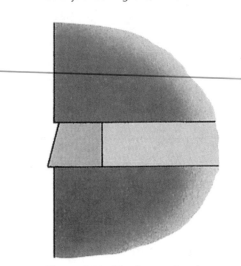

Figure 2.69. *Weather struck and cut pointing.*

Figure 2.70. *50, 75, 100 & 150 mm trowels.*

Figure 2.71. *Cross joint.*

Figure 2.72. *Cutting a cross joint.*

Figure 2.73. *Filling bed joints.*

by the existing brickwork which if too great would prevent complete hydration resulting in a weak crumbly mortar.

Cross joints

Cross joints are filled first. The pointing mortar should be firmly pressed home and compacted with the inset on the left-hand side and the 'cut' on the right-hand side so that every joint looks the same width *(fig 2.71).*

Both left and right-handed bricklayers must inset on the left and cut on the right to prevent the completed wall face having a patchy appearance *(fig 2.72).*

The inset and cut projection must not be exaggerated. In both cases 1 mm is enough.

When the cross joints are completed over about 1 m² of wall surface, the top and bottom 'tails' of mortar should be pressed away.

Bed joints

A trowel with a longer blade is used to apply the mortar to bed joints.

Press the pointing mortar firmly into the joints, insetting just 1 mm at the top and 'cut projecting' the lower edge by the same amount. Sloping or 'weathered' surfaces, by shedding rainwater more readily, are considered to provide better rain resistance than recessed or even flush joints *(fig 2.73).*

Bed joints are cut guided by a pointing rule, a wooden straight edge with spacing blocks to hold it off the surface of the brickwork. Joints may be cut using the point of a towel *(fig 2.74a),* but some bricklayers find that a specially made cutting tool

Specialist pointing gangs are usually engaged for large areas of walling.

Careless pointing can spoil good brickwork and conversely, good pointing can considerably improve 'questionable' facework.

Before pointing begins, loose debris should be removed from the joints with a dry brush and the work 'wetted down' to a damp condition. Wetting down reduces the amount of water sucked from the pointing mortar

Figure 2.74a. *A pointing rule and trowel in use.*

Figure 2.74b. *A pointing rule and Frenchman.*

Figure 2.74c. *A pointing rule and Frenchman in use.*

known as a 'Frenchman' is more manageable (*fig 2.74b,c*).

It is important to cut the mortar bed joints so that they all appear to be the same width.

Timing

Neither cross joints nor bed joints should be cut until the mortar has stiffened a little. This will ensure a clean cut. The absorption of the bricks and weather conditions will determine the timing.

Bricks with a high absorption will allow cutting to be carried out sooner than bricks of low absorption. Similarly, cutting can be carried out sooner in warm drying weather than in cold damp weather.

Figure 2.75. *External angles at corners and reveals.*

Pointing should not be carried out if frost is likely or after a long period of frost if the bricks are still frozen.

External angles

Bed joints should be neatly formed at corners and reveals (*fig 2.75*).

Brushing

At the end of the day a light bristle brush may be used to remove any crumbs of excess mortar left after cutting the

Figure 2.76. *The result of brushing joints too soon.*

joints. Great care should be taken to avoid making brush marks in the soft mortar. It may be advisable to leave brushing the pointing which has been completed late in the day, until the following morning *(fig 2.76)*.

REPOINTING

Before old brickwork is repointed the cause of the deterioration should have been established. It is usually the result of slow erosion over many years but if it is due to, say, sulfate attack on the mortar the cause should be remedied first.

The sequence of operations for repointing old brickwork is virtually the same as for new walling, except that the joints which may be heavily weathered or perished must be cut out first using a hammer, bolster and 'timber dog' with care in order to cause minimal damage to the bricks. It is essential that the recess so formed be left square. The brick edges should be absolutely free of old mortar so that the new mortar *(fig 2.77)* can bond effectively with the bricks.

The recess should be no less than 10 mm and no more than 15 mm deep. If it is too shallow

the mortar may not have a sufficient bond with the bricks and if it is too deep it may be difficult to force the mortar in for the full depth.

At the same time lichen and moss should be removed by careful brushing so as not to damage the bricks.

The brickwork should then be dampened, not soaked, and work should proceed as described for pointing new brickwork.

MORTAR MIXES

Pointing mortars should be 'fatty' and cling to the trowel. This can be achieved by the addition of lime which improves the cohesiveness of mortar, its bond with the bricks and the rain resistance of the brickwork.

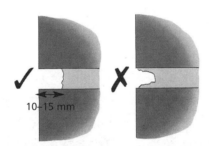

Figure 2.77. *Recessing joints for pointing.*

Under normal conditions of exposure a 1:1:6 cement:lime:sand mix will be specified as appropriate. With calcium silicate bricks a 1:2:9 mix may be required. With very dense bricks only and in situations of extreme exposure the mortar may be 1:¼:3.

In general the mortar should be no stronger than that used in the construction of the wall.[1]

MORTAR BATCHING

Whether the cement, lime and sand are all mixed on site or cement is added to premixed lime:sand, careful measurement for every batch is essential if mortar colour variations and patchy brickwork are to be avoided. It is virtually impossible to achieve satisfactory results if measurement or gauging is done by the shovelful.

Consistent results are also difficult if pigments are added on site. The use of premixed lime:sand for mortars is the only practicable way of producing coloured mortars.

Reference
(1) BS 8000:Part 3:1989 'Workmanship on building sites' Commentary to cl.3.2.2.

KEY POINTS

- Remove lichen and moss from old brickwork before repointing.
- Ensure joints are recessed square and all old mortar and dust is removed before pointing or repointing begins.
- Dampen the wall by wetting down and allow to drain before starting work.
- Press mortar into cross joints first, followed by the bed joints.
- Do not exaggerate the inset or projection of the mortar.
- Cut joints so that all appear the same size.
- Do not brush the finished work too soon.

2.9   BRICKS OF SPECIAL SHAPES AND SIZES

The familiar brick shape has proved to be the most suitable for building and manufacture for over three thousand years, and it will be referred to here as the 'standard' brick.

Even so, from earliest times special shapes have been made to fulfil functional and decorative requirements and today their use has been revived to meet the need for attractive and enriched buildings. For the bricklayer, laying special shapes requires care and attention if the architect's design is to be realised in practice.

THE PURPOSE OF SPECIAL SHAPES

To create shapes in brickwork which would be impossible, unsatisfactory or expensive using only 'standard' bricks.

DEFINITIONS

Bricks of special shapes and sizes
Such bricks are referred to in British Standards as 'Bricks of special shapes and sizes' or more commonly as 'special shapes' or even just 'specials' *(fig 2.78)*.

Standard specials
The term 'standard specials' refers to those shapes and sizes specified in BS 4729. It does **not** imply that manufacturers or suppliers generally hold them in stock.

Non-standard specials
This term describes any bricks of special shapes or sizes **not** specified in BS 4729. They are sometimes referred to as 'purpose made specials' or in everyday speech as 'special specials'.

AVAILABILITY AND STOCKS

Some of the more commonly used standard specials, e.g. single bullnose and cants and plinth headers and stretchers are stocked by suppliers but they will frequently be made to order.

Some special shapes may take longer to produce than 'standard' bricks because they are formed, dried, fired and handled by different processes and in some cases new moulds or extrusion dies will have to be made. This should be taken into account when programming brickwork requirements and placing orders.

STORAGE

All 'specials' take longer to produce and are more expensive than 'standard' bricks and should be carefully and systematically stored and protected from rain in order to reduce damage and wastage and make it easier to find particular types when required.

Money spent on appropriate storage is likely to be justified, particularly as wasted 'specials' can seldom be replaced quickly.

Figure 2.78.

BASIC FUNCTIONS

Special shaped bricks may be grouped according to their most usual function:

- for changes of direction at **angles** other than 90 degrees e.g. 'squints, external and internal angles (doglegs)'.
- for changes in **thickness** vertically, e.g. 'plinth bricks'.
- for **chamfered** and **radiused** corners, e.g. 'cant and bullnose bricks'.
- to **stop**, neatly and effectively, the freestanding end of a run of brickwork or to change from one profile to another, e.g. stops for 'standard', cant or bullnose brick-on-edge cappings.
- to **return** cappings and soldier courses neatly and effectively.
- for **bonding** brickwork without cutting 'standard' bricks, e.g. 'King and Queen closers'.
- for tightly **curved** brickwork, i.e. 'radial bricks'.
- for **arches**. Tapered arch bricks are available for building semi-circular arches with parallel-sided joints between each arch brick.

Notes:
1. Many 'specials', because they have single frogs or have a texture so that the brick has a 'top and bottom', are available in left-hand and right-hand versions, e.g. king closers, cants, bullnoses, squints, angles and certain returns.
2. Because standard and special shape bricks are sometimes formed or fired in different ways the colour or texture may vary slightly. If this is thought to be unacceptable the supplier or manufacturer should be informed immediately and certainly before the bricks are walled in.

SETTING OUT

Squints, external and internal (dogleg) angles

Walls which include angle bricks should be set out to the **face side** as with any facework. The same rules of bonding must be applied at obtuse angles as apply at right-angled corners *(fig 2.79)*. The use of squints instead of external angles gives a smaller face lap but it appears to have proved adequate in practice. Squints have an advantage that being easier to manufacture the angle is often more accurate.

BS 4729 includes a range of angle bricks for internal and external angles of 30°, 45° and 60°. There is a choice of sizes to turn the corner and maintain bond in half lapped or quarter lapped bonding with or without the addition of closers or three-quarter bats.

Plinth bricks

Bonding should be set out so that perpends align vertically

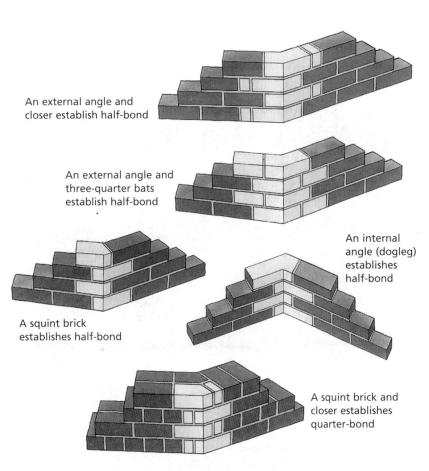

An external angle and closer establish half-bond

An external angle and three-quarter bats establish half-bond

A squint brick establishes half-bond

An internal angle (dogleg) establishes half-bond

A squint brick and closer establishes quarter-bond

Figure 2.79. *Examples of bonding of angle bricks at a quoin.*

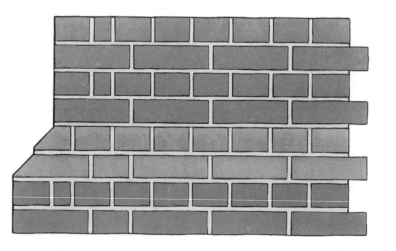

Figure 2.80. *Bonding plinth bricks.*

from the base below the plinth, through the plinth course to the walling above *(fig 2.80)*.

This requires pre-planning by the supervising bricklayer **before** work begins, using pencil and paper to ensure that the correct quoin bricks are used, as a mistake will be permanent and conspicuous.

Pre-planning ensures that any broken bonding is at or near the corner **below** the plinths, then, when the plinth courses have been bedded, the broken bonding at the corner automatically disappears.

Cant and bullnose bricks

Setting out is the same as for walls with normal bricks at the corners but special attention should be paid to plumbing the corners as indicated below.

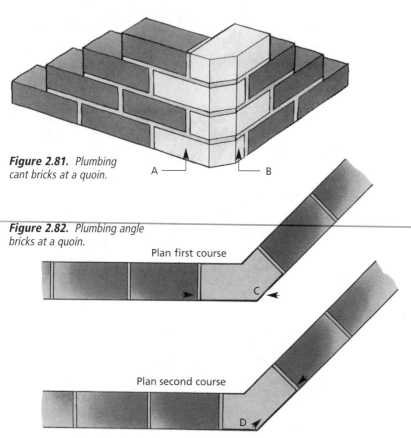

Figure 2.81. *Plumbing cant bricks at a quoin.*

Figure 2.82. *Plumbing angle bricks at a quoin.*

Plan first course

Plan second course

PLUMBING

When plumbing single cant or bullnose bricks, simply tap the quoin headers and stretchers backward or forward until the faces next to the corner are truly vertical at A and B *(fig 2.81)*.

When plumbing squint or external angle quoin bricks they should be tapped along the line of the wall in order to plumb the faces, marked C and D *(fig 2.82)*.

This skill in laying squint and external angle bricks needs to be developed so that the angle

Figure 2.83. *A well plumbed splayed quoin using textured and handed, squint bricks.*

looks truly straight when viewed from below *(fig 2.83)*.

LAYING TO LINE
Special thought must be given to fixing a line and pins when 'running-in' a course of some 'special' bricks *(fig 2.84)*.

It is good practice to consider the most obvious 'sight line' such as the edge or arris of any course of 'specials' which will be immediately apparent to anyone looking at the finished building.

HANDING
Some smooth faced special shapes such as single cants and single bullnoses may be reversed when used for example on either side of an opening *(fig 2.85)*.

This is not so with many textured bricks which initially look and weather differently if laid 'upside down' *(fig 2.86)*. For these, separate right and left handed versions must be ordered and stored carefully for easy and obvious identification, but reference should always be made

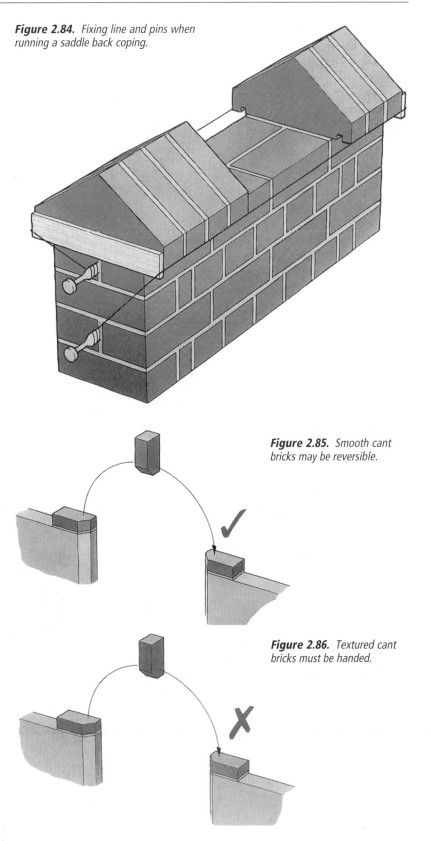

Figure 2.84. *Fixing line and pins when running a saddle back coping.*

Figure 2.85. *Smooth cant bricks may be reversible.*

Figure 2.86. *Textured cant bricks must be handed.*

to BS 4729 to check the correct version before placing an order.

External plinth returns are another example where left and right handed versions are required *(fig 2.87)*.

Right and left handed versions of single frogged specials may be specified for reveals to openings where they have to be laid frog uppermost in order to resist heavy loads from a lintel above. Alternatively, if the appearance of the special allows it to be laid either way up the frogs may be filled with mortar and allowed to set before it is laid frog down in the work.

Some manufacturers stamp 'LH' or 'RH' as appropriate to help identification.

HOW TO IDENTIFY WHICH HAND – LEFT OR RIGHT?

The following method to identify left and right hand versions of specials is used in BS 4729.

When the special is placed on its *normal* bed surface and its equivalent stretcher face viewed, if the modification of the shape is to or at the left hand end of the brick it is a left hand brick. If it is to or at the right hand end it is a right version.

Do not identify right or left hand by brick's intended position in the work.

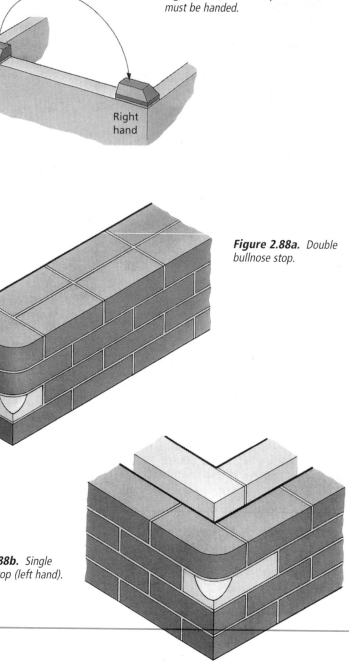

Figure 2.87. *External plinth returns must be handed.*

Left hand

Right hand

Figure 2.88a. *Double bullnose stop.*

Figure 2.88b. *Single bullnose stop (left hand).*

STOP BRICKS

These provide a transition from a special shape to a 'standard' brick (e.g. *fig 2.88a*). Left and right hand versions will be required for some stop bricks (e.g. *fig 2.88b*).

STOP ENDS

Large bullnose and cant bricks (215 × 215 mm or 215 × 159 mm) stop straight runs of cant or bullnose bricks on edge effectively and securely *(fig 2.89)*.

RETURN BRICKS

These allow a special shape to be returned at right angles neatly and securely *(fig 2.90)*.

BONDING BRICKS

One of the essential skills of a bricklayer is the efficient use of hammer and bolster in cutting bats and closers from whole bricks for bonding purposes. However, in order to save time and wastage where large numbers of these are required it is usually possible to obtain them specially made *(fig 2.91)*.

ARCH AND RADIAL BRICKS

The use of these specials for building arches and curved brickwork is covered in sections 5.3, 5.4 and 5.7.

Reference

(1) BS 4729:1990 'Dimensions of bricks of special shapes and sizes'.

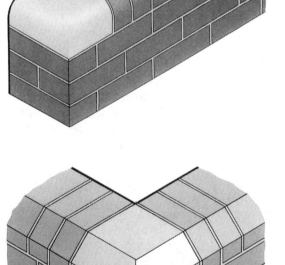

Figure 2.89. *Stop end to double bullnose.*

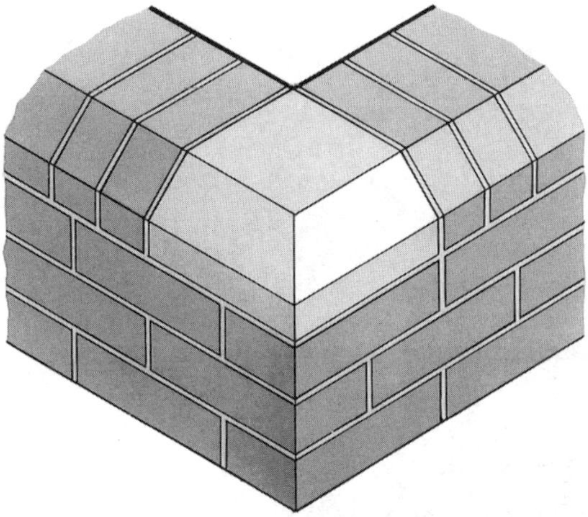

KEY POINTS

- Check if any 'specials' are handed.
- Check whether purpose-made bats and closers are available on site.
- Protect all specials from damage and waste. They may be difficult to replace.
- Take particular care to plumb squint and angle bricks.
- Maintain perpends of face brickwork through plinth courses.
- Consider the 'sight line' when laying a course of projecting 'specials' to the line.

Figure 2.90. *Single cant return.*

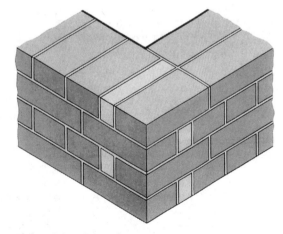

Figure 2.91. *Queen closer.*

3 GOOD PRACTICE

This section deals with some aspects of the work that are not strictly concerned with the manipulative skills involved in the craft, nonetheless they need to be understood and implemented to achieve good results –

working in hot and cold conditions, blending bricks to achieve uniformity of colour.

This section also deals with some specific constructions and highlights the inclusion of

accessories and components generally incorporated into modem brickwork – ties, insulation, DPCs, pipework, manhole covers, etc.

3.1 AVOIDING DAMAGE FROM EXTREMES OF TEMPERATURE

This section is concerned primarily with the prevention of frost damage to brickwork, but makes reference to preventing damage during hot, dry weather.

Preparations for winter building should be made well in advance and equipment and materials procured ready for use.

MORTARS AND FROST DAMAGE
Cements set more slowly at lower temperatures. If mortars freeze before the set is complete, their ultimate adhesion, strength and durability will be reduced

and the brickwork may have to be rebuilt (fig 3.1).

Avoiding frost damage to mortars
Preventing stocks of sand and lime:sand from becoming wet will prevent ice forming within them.

Simply cover them with waterproof, preferably insulated, sheets supported by a light framework arranged to maintain an air space immediately over the stock pile.

Heating of aggregates, bricks and blocks, as has been recommended in the past, does not appear to have proved practicable.

Tubs of retarded mortar should be covered to prevent the mortar freezing.

Bricklaying should stop when the air temperature falls to 3°C (fig 3.2) unless the temperature of the mortar can be maintained at a minimum of 4°C until it has

Figure 3.1. The result of frost damage to unset mortar.

Figure 3.2. Checking the air temperature from a conveniently placed thermometer.

hardened, as recommended in the Masonry Code of Practice[1].

Bricklaying should resume only when the air temperature reaches 1°C and is still rising – but only if the bricks are not frozen. To test, brush cold water on to the faces of some of the bricks with a paint brush; if the bricks are frozen the water will turn to ice.

A more direct method uses a 'spear' thermometer to check the temperature of a trial mortar bed left to cool for a suitable time (fig 3.3).

Brickwork can be protected from wind by heavy gauge, reinforced polythene sheeting secured to the scaffolding. Alternatively, work areas and materials can be enclosed in heated lightweight structures (fig 3.4).

If frost is likely before the mortar has set, the brickwork should be protected by waterproof insulation (see Section 1.2 'Protection of newly built brickwork').

Protection from frost should be left in place for about 7 days. If there is any doubt about the condition of the mortar, heat selected small areas to above freezing point. If the mortar is found to be soft, it has not set, and the brickwork may have to be rebuilt.

Take care not to stain the brickwork. A radiant heater may be suitable.

'Anti-freeze' agents

Accelerators, although successfully used in concrete, are not recommended for masonry mortars. This is because the additional heat liberated from the accelerated hydration of small amounts of cement will be quickly conducted away by the bricks and blocks which have a relatively high thermal capacity.

Admixtures containing calcium chloride should never be used as they may lead to dampness and corrosion of embedded metals like wall ties and reinforcement.

As there is no UK experience of admixtures which reduce the freezing point of mortars being mixed, they are not recommended. Some may adversely affect the hydration of cements.

The frost resistance of mortars during the setting can be improved by adding air-entraining admixtures to the mix. However, their use may reduce both the mortar strength and adhesion and should be added only with the permission of the architect or engineer. They should comply with BS 4887:Part 1[2] and the manufacturer's instructions should be strictly followed.

Ready-mixed lime:sand for mortars and ready-to-use retarded mortars supplied to sites may contain the optimum quantity of air-entraining admixture. Unauthorised extra admixtures should not be added.

Protecting bricks from frost

Moderately frost resistant (M) clay bricks are at risk from frost attack when saturated and

Figure 3.3. A spear thermometer in use.

Figure 3.4. External and internal views of a temporary enclosure.

should be protected, whether in the packs, stacked on the scaffolding or in the wall (see Section 6.2 'Frost attack and frost resistance'). But even frost resistant (F) bricks should be protected from saturation as, if the water content freezes, the bricks will be unusable until they thaw.

Protection of the mixing area

The temperature in the area of the mixing plant and materials may be kept higher and working conditions made more comfortable by the erection of simple wind screens.

BRICKLAYING IN WARM DRYING WEATHER

Bricks of high water absorption usually have a high suction rate. They rapidly absorb moisture from mortars which not only reduces their workability but may seriously reduce adhesion.

Mortar should be laid in short lengths to limit the loss of water before the bricks are laid.

It is considered that the use of lime in mortars helps to improve their bond with clay bricks that have a high water absorption (see page 77, Section 4.1 'Mortars').

Newly built brickwork should be protected from drying out too rapidly as this may result in a reduction in strength (see page 5, Section 1.2 'Protection of newly built brickwork').

Clay bricks

Adhesion of mortars to highly porous clay bricks may also be improved by 'docking' them, but they must not be overwetted as this may lead to

'floating' on the mortar bed and to efflorescence and staining of the facework. Bricks of low water absorption should not be wetted.

Calcium silicate bricks

Easier laying and better adhesion is achieved by adjusting the consistency of the mortar or briefly docking, not soaking, the bricks just before laying.

WEATHER FORECASTS FROM THE METEROLOGICAL OFFICE

There are a number of forecast services available for builders via fax, phone or the Internet.

Fax – MetFAX is a premium rate fax service, giving five day regional weather forecasts. An index page listing the areas and fax numbers can be obtained by dialling 09060 100 400.

Phone – MetCALL DIRECT gives you the opportunity to speak directly to a forecaster day or night. To access this service call 08700 767 828. Payment is by

credit card on-line. For more details on past weather information for planning, contract disputes and project overruns call 08700 767 876. **Internet** – a complete range of forecast services is available on MetWEB at www.metoffice.gov.uk. To subscribe or to obtain further information call 08700 750 077.

Full details of The Met. Office's services to builders can be found on their FREE index card which you can get by calling their helpline on 08700 750 075 or by e-mail: metfax@meto.gov.uk.

At the time of writing (November 1999), 09060 numbers are charged at £1 per minute at all times.

References

(1) BS 5628:Part 3:1985 'Code of Practice for use of Masonry' (cl. 35.3 'Work in cold conditions').
(2) BS 4877:Part 1:1986 'Specification for air-entraining plasticizer admixtures'.

KEY POINTS

- Anticipate and prepare for cold weather.
- Cements set more slowly in cold weather.
- Brickwork may need rebuilding if mortars are frozen before the set is complete.
- 'Anti-freeze' admixtures are of little value in masonry mortars.
- Never use calcium chloride.
- High absorption clay bricks may need docking in hot, dry weather.
- With calcium silicate bricks, preferably adjust the mortar consistency.
- Ensure that a robust, reliable thermometer is available on site.
- Use a weather prediction service to anticipate need for precautions.

3.2   BLENDING FACING BRICKS ON SITE

Packs of facing bricks are blended or mixed as they are loaded-out to bricklayers in order to minimise the effect of slight but inevitable variations in colour and size on the appearance of finished brickwork.

THE CAUSES OF COLOUR AND SIZE VARIATIONS – MANUFACTURING CONTROLS

Variations in raw materials
Bricks are made mainly from natural materials deposited at different times over millions of years, often in layers having very different physical qualities.

Clay deposits vary considerably having been formed by the weathering of rocks; the depositing of sediments by rivers and lakes and finally by geological upheavals and further weathering.

Winning, processing and forming clay
Although, during the winning and stockpiling of clay, the various materials are mixed as homogeneously as is practicable (fig 3.5), slight variations persist, particularly between stockpiles produced at different times.

Raw materials from stockpiles are crushed and finely ground and other materials may be added. The final mixture may be moulded or pressed or extruded and wire-cut to shape before being dried ready for firing. *(see Section 6.10 'Brick manufacture')*

Firing clay bricks
Modern tunnel kilns are continuously monitored and controlled to minimise differences in firing temperatures from one part of a kiln to another, and from time to time *(fig 3.6)*. Even so, some slight variations will remain, particularly between the top and bottom of a kiln. Larger variations occur with some other methods of firing e.g. in clamps and intermittent kilns.

Calcium silicate (sandlime and flintlime) and concrete bricks
Calcium silicate bricks consist of aggregates, lime and pigments processed in high pressure steam autoclaves. Concrete bricks consist of aggregates, cement and pigments cured similarly to other concrete products. Both types of bricks can vary in colour and size and although the same principles for blending apply as for clay bricks it is advisable to follow the manufacturer's particular recommendations in this regard.

Control of variations
During all manufacturing processes, the use of modern and traditional skills, sophisticated plant and quality control techniques minimise variations in colour and dimensional tolerances. The risk of these variations adversely affecting the appearance of

Figure 3.5. *Stratified quarry face and stockpiles layered for weathering.*

Figure 3.6. *Computer managed kiln control.*

finished brickwork can be minimised by supervisors, bricklayers and labourers following the recommendations in this section.

THE NEED FOR SITE BLENDING

Variant bricks within a load can, if grouped together, result in unacceptable colour patches, patterns or bands in brickwork *(fig 3.7)*.

Blending disperses the few variants among the many typical bricks to achieve brickwork free of patches.

The need to blend bricks that are intended to be uniform in colour (e.g. smooth red bricks) is generally and readily understood, but it is not always understood that it is **absolutely essential to blend multicoloured bricks** to avoid unwanted patterns caused by bricks of one particular colour being grouped together. *(see Section 6.6 'Appearance')*

Because variations in raw materials and firing of clay bricks can result in variations of size as well as colour, blending of bricks will also help bricklayers to maintain regular widths of cross-joints when setting-out facework. *(see Sections 6.5 'Allowing for variations in brick sizes', and 2.1 'Setting-out facework – stretcher half-bond')*

There is likely to be greater variation between different loads than within a single load. The blending techniques described in this section apply to the latter case. Minimising the adverse effects of variations between loads *(fig 3.9)* should be tackled by those who specify and order bricks. They should, as soon as possible, alert the manufacturers to the need to supply bricks of a uniform appearance over a long period of time. It is not realistic to

Figure 3.7. *Failure to blend loads when loading out.*

Figure 3.8. *Failure to ensure uniformity between loads.*

expect manufacturers to anticipate this.

Modern methods of handling and transporting bricks

In the past, bricks were inevitably blended when they were handled, six or eight times, from kiln to stockpiles; from stockpiles to lorries; from lorries to site stacks and finally loaded-out for the bricklayers. These four processes were akin to shuffling a pack of playing cards *(fig 3.9)*.

Today, bricks from kilns are formed into strapped packs of some four hundred and moved mechanically to site stacks on the ground, loading platforms on floors of buildings. Unless packs are blended as they are loaded-out to bricklayers there is a risk of similar coloured bricks being grouped in the wall causing patchy brickwork. Blending also mixes bricks having different size tolerances, helping bricklayers to maintain greater regularity of cross-joints when setting out facework. *(see Section 2.1 'Setting-out facework – stretcher half-bond')*

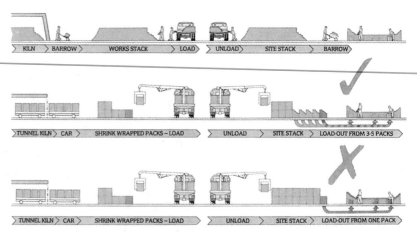

Figure 3.9. *The need for site blending.*

THE TECHNIQUES OF BLENDING ON SITE

Bricks should be loaded-out preferably from **four or five packs but from an absolute minimum of three**. It is usually advisable to draw from packs in **vertical rather than horizontal slices** as the position of bricks in packs tends to reflect their position in the kiln *(fig 3.10)*. This is particularly true of tunnel kilns. *(see Section 6.6 'Appearance')*

If in doubt ask the manufacturers for their advice.

If bricklaying gangs understand the reasons for loading-out in this way there should be no problems of colour patches or banding when working the first lift.

However, to service bricklayers on upper lifts of scaffolding, loading bays need to be large enough for at least three or four packs. If loading bays are not large enough some difficulty may be experienced in avoiding colour banding.

Loading bays up to about 7 m high must have hinged baffler/guard rails for access by fork lift truck or conveyor.

Above 7 m, packs of bricks can be lowered directly on loading bays by tower cranes.

An alternative method is to load barrows from a number of packs at ground level and transport them vertically by hoist or conveyor, but special care must be taken not to chip the bricks when handling them in this way.

Blending on site alone is not effective in eliminating colour variations between loads delivered to a site over a long period of time. The manufacturer should be warned prior to work commencing. It is important that, if practicable, blending takes place between loads of bricks, as well as between packs, especially if loads are from different manufacturing batches.

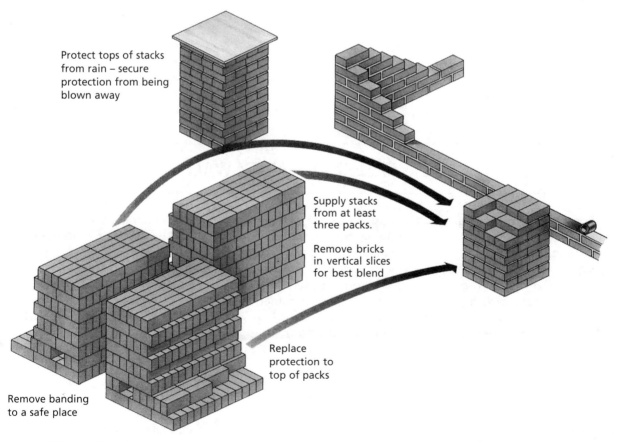

Protect tops of stacks from rain – secure protection from being blown away

Supply stacks from at least three packs.

Remove bricks in vertical slices for best blend

Replace protection to top of packs

Remove banding to a safe place

Figure 3.10. *Loading-out from a number of packs.*

BLENDING IN FACTORIES
Some manufacturers are able to blend some bricks in the factory before dispatch, but further blending on site will usually be advisable. Manufacturers should be consulted.

SOME OTHER ASPECTS
Although handmade/soft-mud facing bricks have two usable faces, the overall colour may be different on opposite faces.

Moulded bricks should be laid so that they smile. *(see Section 6.6 'Appearance')*

Extruded wire-cut bricks often have a directional texture such that they should be laid in the same orientation, i.e. 'the right way up'.

But, blending of bricks will be to little avail if a consistent mortar colour is not achieved. *(see Sections 1.2 'Protection of newly built brickwork', 1.3 'Handling, storage and*

protection of materials', 4.1 'Mortars', 6.6 'Appearance')

CONCLUSION
Modern methods of manufacture, handling, packaging and particularly the virtual elimination of laborious hand loading has helped to minimise rising costs. Mechanical handling on site has not only

similarly reduced hard labour but facilitates the reduction of wastage from breakages and chipping.

But these modern methods have increased the need for deliberate on-site blending if colour patches and banding are not to leave permanent scars as evidence of a lack of understanding or care and attention by all concerned.

KEY POINTS

- Consult brickmakers about their recommendations for site blending.
- Explain to bricklayers and labourers the reasons for blending, before they begin work on site.
- Ensure that scaffold loading bays are large enough and strong enough to carry a sufficient number of packs.
- Check loads when delivered to site and when broken down for

loading-out so that problems can be identified before the bricks are walled.
- Constantly check that hod carriers are drawing vertical slices preferably from five packs when loading-out.
- Look out constantly for noticeable patches or bands of colour variation at bricklaying level and remedy immediately whilst mortar is green.

3.3   EXTERNAL CAVITY WALLS

There is more to building modern cavity walls than bedding bricks in mortar.

Simple cavity walls began to supersede solid walls *(fig 3.11a)* over fifty years ago because they were more rain resistant. This superiority depends increasingly on careful, detailed design, specification and workmanship as cavity walls become more complex to meet more exacting requirements *(fig 3.11b)*.

Users grow less tolerant of rain penetration; lightly loaded cavity walls require more provision to prevent cracking by movement than do solid walls restrained by roof and floor loads; the durability of wall ties has proved critical in maintaining structural stability; finally, cavity insulation has been widely adopted. Care and attention, based on an understanding of the way cavity walls both succeed and

fail, will minimise expensive and disruptive maintenance and repairs.

THE SCOPE OF THIS SECTION
This section describes brickwork operations, e.g. '12. Building-in ties', and which functional requirements, they mainly affect, e.g. 'Rain resistance'. See table 3.1. Reference is made throughout to detailed descriptions in other sections.

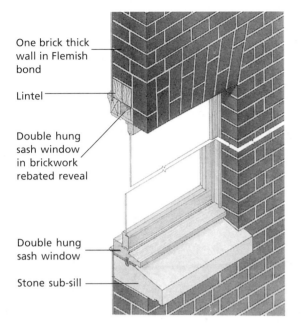

One brick thick
wall in Flemish
bond

Lintel

Double hung
sash window
in brickwork
rebated reveal

Double hung
sash window

Stone sub-sill

Figure 3.11a. *A typical external traditional solid wall.*

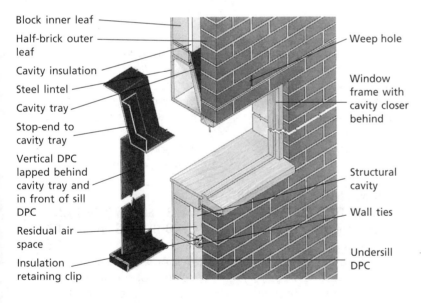

Block inner leaf

Half-brick outer
leaf

Cavity insulation

Steel lintel

Cavity tray

Stop-end to
cavity tray

Vertical DPC
lapped behind
cavity tray and
in front of sill
DPC

Residual air
space

Insulation
retaining clip

Weep hole

Window
frame with
cavity closer
behind

Structural
cavity

Wall ties

Undersill
DPC

Figure 3.11b. *A typical external modern cavity wall.*

1. CHECKING MATERIALS ON DELIVERY

Check that all deliveries are complete and undamaged before accepting them. **In particular check that**:

- **Facing bricks** are as specified; minor surface blemishes are acceptable; *(see Section 1.1 'References and sample panels')*; bricks have been gauged and

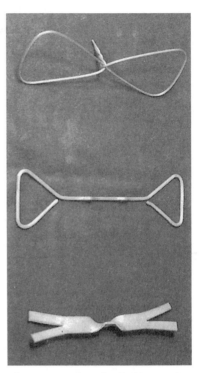

Figure 3.12. *Butterfly, double triangle and vertical twist ties.*

selected for close tolerances, if so specified, for use e.g. in narrow piers, cappings, soldier courses and similar features. *The wrong bricks can result in patchiness and irregular brickwork features.*

- **Concrete blocks** are of the specified type and thickness. *The wrong blocks may have inadequate strength or thermal insulation.*

- **Wall ties** *(fig 3.12)* are of the specified type and length; plastic debonding sleeves are included if required for some ties to allow movement; suitable clips are provided for ties retaining cavity insulation. *The wrong ties can result in unstable walls, cracking, misplaced insulation and rain penetration.*

TABLE 3.1   Functional requirements of cavity walls

	Strength/stability	Rain resistance	Appearance	Movement	Durability	Thermal insulation
1. Checking materials	•	•	•	•	•	•
2. Handle, store, protect	•	•	•	•	•	•
3. Set-out and build facework			•			
4. Cleanliness, protection			•		•	•
5. Mortar mixing	•		•	•	•	
6. Forming cavities		•				
7. Building blockwork	•			•		•
8. Raising two leaves	•		(and safety)			
9. Jointing	•	•	•		•	
10. Parapet walls	•	•	•	•	•	•
11. Fixing window frames	•	•				
12. Building-in ties	•			•		•
13. Building-in DPCs	•	•	•	•	•	•
14. Building-in insulation		•				•
15. Movement joints	•	•	•	•		
16. Support systems	•	•	•	•		
17. Lintels	•					

- **Mortar materials** e.g. cement, lime, sand, pre-mixed coarse stuff and ready-to-use retarded mortars are as specified; they are from constant sources if for use in facework. *The wrong materials can lead to weakened and less durable mortars, variations in mortar colour and patchy looking brickwork.*
- **DPCs** are of the specified type and width; the appropriate adhesives are included for sealing laps at joints; that any preformed units have the right dimensions. *The wrong DPCs may be too narrow to be built-in effectively or have inadequate bond with mortar. The wrong adhesives may result in inadequate sealing at laps, allowing rain penetration.*

- **Cavity insulation materials** are of the correct type, size and thickness. *The wrong materials can lead to inadequate thermal insulation and rain penetration.*

2. HANDLING, STORING AND PROTECTING MATERIALS

Handle, store and protect all materials to avoid damage by impact, abrasion, excessive loads, rain, ground water, heat, cold and contamination by other materials. *Damaged bricks and mortar materials can spoil the appearance of brickwork as well as reducing its strength and durability. Damaged insulation and DPCs can lead to rain penetration and reduced thermal insulation. Damaged lintels and ties can result in*

weakened walls. (see Section 1.3 'Handling, storage and protection of materials')

Load-out bricks preferably from five packs to avoid colour patchiness or banding of brickwork. *(see Section 3.2 'Blending facing bricks on site')*

3. SETTING-OUT AND BUILDING FACEWORK

Set-out facework dry at ground level. Agree with supervisor the positions of openings and any broken or reverse bond; establish and maintain perpends; use gauge rods for regular coursing. *Facing brickwork of distinction is achieved by careful preparation before a brick is laid and the continuing exercise of basic bricklaying skills with care and attention. (see Sections 2.1 to 2.9)*

Figure 3.13. *Some points for care when building cavity walls.*

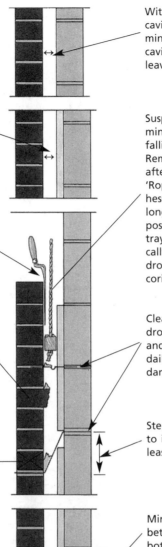

With full-fill or no cavity insulation – a minimum 50 mm cavity between leaves

With partial-fill cavity insulation – a recommended minimum 50 mm residual air space

Clean excess mortar from cavity side of both leaves, especially when building-in full-fill insulation

Avoid protrusions in cavity. Snapped headers, if required, should be purpose-made or accurately and cleanly cut

Immediately above DPC trays leave cross-joints open as weep holes at not more than 1 m centres but with at least two above any opening. Keep them clear of debris. Fit filtration plugs if required

Minimum 150 mm between DPC and ground level

Weep holes every fourth cross-joint

Suspend lath to minimise mortar falling down cavity. Remove and clean after six courses. 'Ropes' of twisted hessian, about 3 m long, may be positioned above trays and, periodically, carefully drown out through coring holes

Clean mortar droppings from ties and cavity trays daily. Do not damage trays

Step cavity tray up to inner leaf at least 150 mm

Minimum 150 mm between DPC and bottom of cavity

Leave shallow space at ground level for unavoidable mortar droppings

4. CLEANLINESS AND PROTECTION OF FACEWORK

Keep facework free from mortar smears and splashes; protect projecting features such as plinths and string courses from mortar droppings; protect brick sills, reveals and soffits from mechanical damage; turn back scaffold boards nearest to facework if rain is likely; protect newly built brickwork from rain and frost, particularly overnight. *Finished brickwork can be permanently scarred by carelessness and if not* *protected from damage by subsequent building operations and the weather*. *(see Sections 1.2 'Protection of newly built brickwork'; 2.6 'Keeping brickwork clean')*

5. BATCHING AND MIXING MORTAR

Batch mortar materials accurately, and be aware that different mixes may be specified for particular locations, e.g. below DPC level, in parapet walls, copings, cappings, brick sills and inner leaves.

When site-mixing mortars use a method recommended in section 4.1. *Incorrectly batched or mixed mortars can mar the appearance of brickwork and reduce its strength, durability and resistance to cracking.* *(see Sections 4.1 'Mortars'; 6.2 'Frost attack and frost resistance'; 6.3 'Sulfate attack on mortars'; 6.4 'Durability of brickwork'; 6.6 'Appearance')*

6. FORMING CAVITIES

Observance of the good practice points noted in figure 3.13 will minimise the risk of rain penetration and costly remedial work.

7. BLOCKWORK INNER LEAVES

Do not mix bricks and blocks or different blocks in the inner leaf. *This can reduce wall strength and thermal insulation and cause pattern staining.*

Chases for services should not exceed the dimensions shown in *fig 3.14*. Sloping chases are not recommended because the difficulty of establishing their

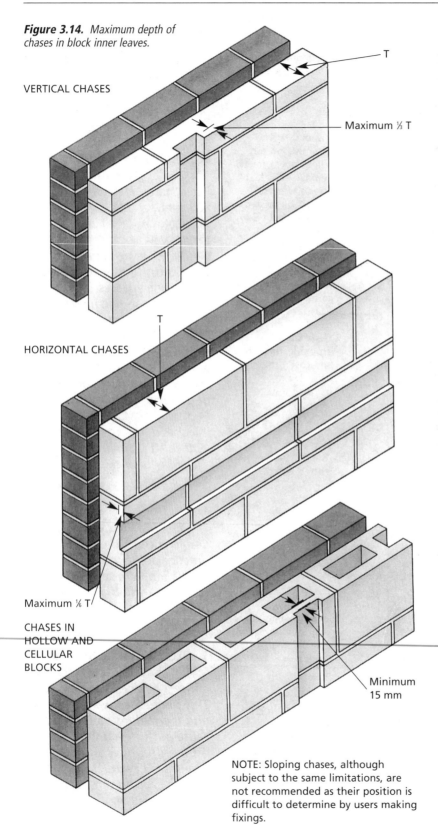

Figure 3.14. *Maximum depth of chases in block inner leaves.*

VERTICAL CHASES

T

Maximum ⅓ T

HORIZONTAL CHASES

T

Maximum ⅙ T

CHASES IN HOLLOW AND CELLULAR BLOCKS

Minimum 15 mm

NOTE: Sloping chases, although subject to the same limitations, are not recommended as their position is difficult to determine by users making fixings.

positions in finished walls can be hazardous for users making fixings. If in doubt ask. *The indiscriminate cutting of chases can seriously weaken masonry walls*.

Internal leaves are generally bonded to internal walls for stability. But check the detailed drawings; special provision may be required to accommodate movement and control shrinkage cracking. *(see Section 6.11 'Building blockwork inner leaves, walls and partitions')*

8. DIFFERENCE IN HEIGHT WHEN RAISING TWO LEAVES *(fig 3.15)*

Both leaves should be raised in one operation. At no time should the difference in height be more than six courses of blockwork (approximately 1350 mm) except when using vertical twist ties when the difference should be no more than two block courses (approximately 450 mm). *Unsupported, newly-built single leaves are liable to be blown over in strong winds*.

These recommendations are summarised in Table 3.2 and are based on clause 32.11.1 of BS 5628:Part 3[1].

These recommendations may be discounted when the site supervising engineer makes allowance for the reduced stability of the single inner leaf in resisting wind load (e.g. by providing temporary support or shelter) and special two-part ties *(fig 3.16)* are specified to overcome misalignment of bed joints and the potential danger caused by the sharp ends of ties projecting from the inner leaf while the outer leaf awaits completion.

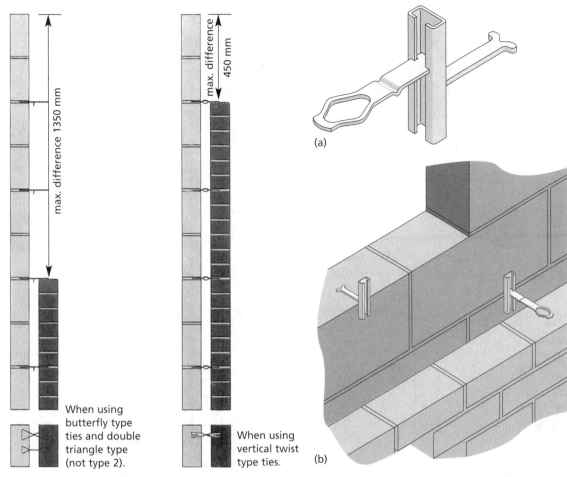

Figure 3.15. *Maximum differences in height when raising two leaves.*

Figure 3.16. *Two-part stainless steel tie (a) and shown in position (b). Image kindly supplied by Ancon CCL.*

TABLE 3.2 Maximum differences in height of both leaves of a cavity wall during construction			
Type of tie		Maximum difference in height of the two leaves	
Shape name to BS 1243 (2)	Type no. to DD 140 Pt 2 (3)	Block courses	mm approx
Butterfly	4, 5 & 6	6 courses	1350 mm
Double triangle	2	6 courses	1350 mm
Vertical twist	1	2 courses	450 mm

NOTE: Unsupported, newly-built single leaves are liable to be blown over in strong winds. The more onerous limitations on differences in height when using 'vertical twist ties' is to minimise discrepancies between the level of bed joints between each leaf. Such discrepancies may cause bricklayers to force the free ends of ties, built into the higher leaf, up or down to suit the level of joints in the rising second leaf. This may disrupt the bed joints and masonry units as the ties are too stiff to bend. In addition, the exposed ends of vertical twist ties can cause injuries to the body and particularly the eyes.

9. FORMING AND FINISHING MORTAR JOINTS

Fill all bed and cross joints solidly *(fig 3.17). Partial filling, particularly of cross-joints, is one of the main causes of increased rain penetration through the outer leaf. (see Section 6.7 'Rain resistance of cavity walls'.) Deep furrowing of bed joints reduces the load-bearing capacity of brickwork and blockwork leaves.*

Unless otherwise instructed, iron all joints externally in the brick outer leaf, e.g. as 'struck' or 'bucket handle'. *This*

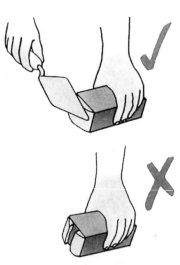

Figure 3.17. *Cross joints. Solid are rain resistant – tipped and tailed are not.*

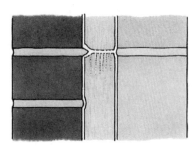

Figure 3.18. *Mortar joint not cut flush leads to rain penetration.*

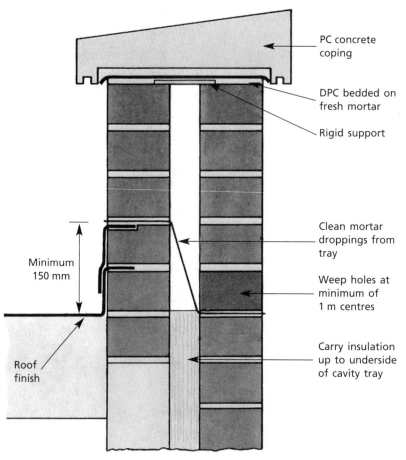

PC concrete coping

DPC bedded on fresh mortar

Rigid support

Clean mortar droppings from tray

Weep holes at minimum of 1 m centres

Carry insulation up to underside of cavity tray

Minimum 150 mm

Roof finish

Figure 3.19. *A typical cavity parapet wall.*

increases the rain resistance of the brickwork.

Do not recess joints in brickwork that may become excessively wet. *This can lead to increased rain penetration of the outer leaf and possible frost damage.*

The jointing techniques of all bricklayers on a job should be co-ordinated. *This is essential in order to maintain a uniform appearance. (see Sections 2.7 'Finishing mortar joints' and 2.8 'Pointing and repointing')*

Clean excess mortar from the back of the outer leaf as the work proceeds. *Excess mortar*

protruding from the joints can cause rain penetration through the horizontal joints between full-fill cavity batts (fig 3.18) (see Section 4.4 'Insulated cavity walls')

10. PARAPET WALLS
Pay special attention to building-in DPC trays and flashings and DPCs under copings and cappings *(fig 3.19)*.

Parapets are the most exposed parts of cavity walls and failures can result in rain penetration, staining and deterioration. (see Section 5.2 'Cavity parapet walls')

11. OPENINGS FOR WINDOW AND DOOR FRAMES
Fix built-in frames with approved cramps, plugs and fixings. Do not use timber plugs which can rot in outer leaves. Position and fix DPCs as described under '*13. Building-in DPCs*'.

Some types of proprietary cavity closers are permitted by building control authorities. Check if in doubt.

12. BUILDING-IN TIES
Build-in the type and length of cavity wall ties specified, not less than 50 mm into each leaf, and **not** sloping down to the inner

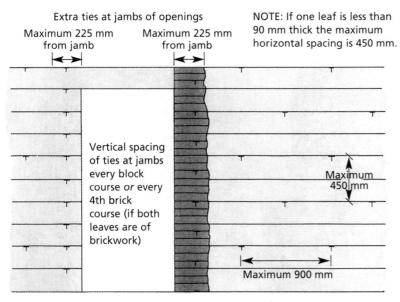

Extra ties at jambs of openings

Maximum 225 mm
from jamb

Maximum 225 mm
from jamb

NOTE: If one leaf is less than
90 mm thick the maximum
horizontal spacing is 450 mm.

Vertical spacing
of ties at jambs
every block
course *or* every
4th brick
course (if both
leaves are of
brickwork)

Maximum
450 mm

Maximum 900 mm

Figure 3.20. *Cavity wall tie spacing.*

leaf. Position any drip in the centre of the cavity. Maintain the specified spacings, taking particular care to increase the number of ties at openings and movement joints *(fig 3.20)*. Keep all ties free of mortar droppings. Ties that are merely pushed into bed joints have little pull-out strength.

Special ties, intended to restrain movement in one direction but allow it in another, must be built-in with care and an understanding of how they work.

The correct ties properly built-in are essential to give the cavity wall sufficient strength and stability, maximum rain resistance, support and restraint to any cavity thermal insulation and to minimise the risk of cracking. (see Sections 4.2 'Ties in cavity walls', 4.4 'Insulated cavity walls' and 4.5 'Vertical movement joints')

13. BUILDING-IN DPCS

(see Section 4.3 'Damp-proof courses'.) Bed only specified DPCs, preferably in a single length, on fresh mortar. Lap unavoidable joints by a minimum of 100 mm. Bond laps in cavity trays with an adhesive recommended by the manufacturer. *Lapping without bonding is acceptable only against rising damp e.g. in ground-level DPCs.*

DPCS immediately above ground-level
Bed DPCs on fresh mortar at least 150 mm above ground or paving level. *This minimises the risk of rain splashing up from hard surfaces and of top soil being placed above the DPC.*

Do not allow DPCs to project into the cavity. *They form ledges on which mortar accumulates and bridges the cavity.*

Project DPCs by 5 mm from or keep flush with the facework. *Mortar and bricks can spall as*

DPCs positioned behind the face and pointed over, compress under load.

If two courses of DPC bricks are specified they must be bedded in a designation (i) e.g. 1:¼:3 mortar to resist rising damp.

Vertical DPCs
At openings, build-in vertical DPCs to protect door and window frames from brickwork that may become wet. Project the DPCs beyond the cavity closer and into the cavity by at least 25 mm but preferably 50 mm. If preformed cavity closers are specified, take care at the junction with DPCs.

Lap vertical DPCs behind DPC trays at lintel level and in front of DPCs at sill level. Any joints must be lapped and sealed *(fig 3.21)*. *This will shed water draining down the inner face of the outer leaf to the outer face through weep holes rather than into the cavities and insulation.*

Cavity trays
Project cavity trays 5 mm or keep flush with the outer face. Step them up at least 150 mm; build into the inner leaf and fix stop ends. Clear mortar droppings from cavity trays, taking care not to damage them *(fig 3.21)*. *Cavity trays are intended to collect water from the cavity and drain it to the outside through weep holes. Gaps, perforations or lack of effective stop ends can allow water to reach lintels, frames and thermal insulation.*

DPCs under copings and cappings
At the top of cavity parapet walls bed DPCs on a rigid bridge over the cavity and bed copings or cappings on the DPC in one operation to maximise the bond

Figure 3.21. *Cavity trays, vertical and sill DPCs.*

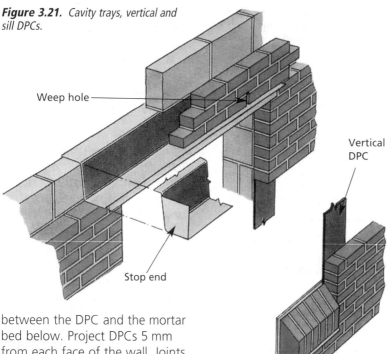

Weep hole

Vertical DPC

Stop end

Figure 3.22. *Protect cavity insulation from mortar droppings.*

between the DPC and the mortar bed below. Project DPCs 5 mm from each face of the wall. Joints must be lapped and sealed.

Failure of these DPCs can lead to the saturation, staining and frost failure of the bricks and mortar and the dislodgement of the coping or capping. (see Sections 5.1 'Copings and cappings'; 5.2 'Cavity parapet walls')

DPCs under sills
Build-in DPCs under sills that are not impervious to water or are jointed. Turn them up at the back if the sill is in contact with any part of the inner leaf.

These DPCs are to prevent water, that permeates through a sill or between joints which eventually crack, from saturating the brickwork below or being transferred to the inner leaf.

14. BUILDING-IN CAVITY INSULATION
Build-in full-fill batts supported on wall ties; batts should fill the cavity

and fit close together with no mortar droppings between them; cut neat slits where necessary to fit batts over ties. Fix partial-fill cavity insulation with special ties, normally back to the inner leaf, leaving the specified residual air space behind the outer leaf. Protect the top of cavity insulation from mortar droppings *(fig 3.22)*. *Incorrectly fitted insulation or mortar droppings between batts or slabs can lead to rain penetration and reduced insulation values. (see Section 4.4 'Insulated cavity walls')*

15. MOVEMENT JOINTS
Keep vertical movement joints clear of mortar, use the specified filler, usually flexible cellular polyethylene, cellular polyurethane or foam rubber but

never hemp, fibre board or cork which are insufficiently compressible for joints in clay brickwork which expands. Apply the specified sealant to the correct depth in accordance with the manufacturer's instructions *(fig 3.23)*. Build-in extra wall ties at movement joints as specified. *(see Sections 4.5 'Vertical movement joints'; 4.7 'Brickwork on metal support systems')*

Extra movement joints will be required in parapet walls.

16. BUILDING CAVITY WALLS ON METAL SUPPORT SYSTEMS
Build with special care those cavity walls in which the outer leaf is supported on metal angles or brackets *(fig 3.24)*. Provide a

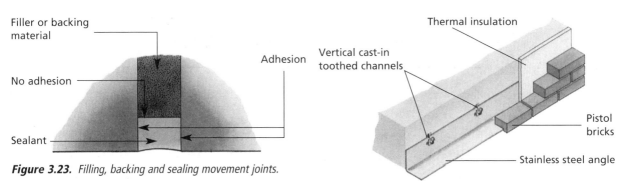

Figure 3.23. *Filling, backing and sealing movement joints.*

Figure 32.4. *A typical continuous angle support system. (To aid clarity a cavity tray has not been shown)*

horizontal movement joint between the underside of the support and the top of any brickwork leaf immediately below; tie the top of the panels back to the structural frame as specified or recommended by the manufacturer of the support system. At least two-thirds of the thickness of the brick outer leaf should bear on the supporting system. Pay particular attention to the building-in of DPC trays. *Failure to follow recommendations can result in reduced strength and stability, rain penetration and a poor appearance. (see Section 4.7 'Brickwork on metal support systems')*

17. LINTELS

Build-in lintels with adequate bearings as specified or recommended by the manufacturer but never less than 100 mm. Bed them on mortar on full blocks, not short lengths of cut blocks. *(see Section 6.11 'Blockwork inner leaves, walls and partitions')*

KEY POINTS

- In all matters described in this section take particular care with cavity walls which will be exposed to considerable wetting from wind-driven rain.
- Check correctness and condition of all materials and components on delivery.
- Handle and store all materials and components to avoid damage and deterioration.
- Raise both leaves together, keeping differences in height to recommendations.
- Build-in all DPCs as recommended with great care.
- Take particular care with DPCs and flashings in cavity parapet walls.
- Maintain specified cavity widths.
- Clean excess mortar from cavity faces.
- Avoid protrusions in cavities from cut bricks and DPCs.

- Minimise mortar droppings in cavities.
- Clear mortar droppings from ties, cavity trays and bottom of cavities daily, avoiding damage to cavity trays.
- Build-in only specified wall ties, correctly positioned and spaced.
- Build-in ties solidly by 50 mm into each leaf, level or sloping down to outer leaf, drips in centre of cavity pointing down.
- Build-in various types of bricks and blocks only in positions specified.
- Cut chases in inner leaf only as instructed or to recommendations.
- Leave open cross-joints for weep holes as specified and keep clear.
- Build-in thermal insulation as recommended with great care to avoid causing rain penetration.

References
(1) BS 5628 Part 3:1985 'Code of Practice for Use of Masonry.'
(2) BS 1243:1978 'Specification for metal ties for cavity wall construction.'
(3) DD140:Part 2:1987 'Recommendations for design of wall ties.'

Further reading
BRE Defect Action Sheet (Design). DAS 12. December 1982 'Cavity trays in external walls: preventing water penetration.'
BRE Defect Action Sheet (Site). DAS 17. February 1983 'External masonry walls insulated with mineral fibre cavity-width batts.'
BRE Defect Action Sheet (Site). DAS 116. June 1988 'External masonry cavity walls: wall ties – installation.
BS 8000:Part3:1989. 'Workmanship on building sites – Code of practice for masonry.'
Brick Development Association: Good Practice Note 1. 'Cavity insulated walls.'

3.4 FROG UP OR FROG DOWN?

A frog is a depression formed in a bed face of a brick.

Most bricklayers will have been instructed at some time that bricks must be laid with frogs up and filled with mortar. At other times they may have been advised that bricks may be laid either way.

Both the instructions and the advice can be correct depending on circumstances. This section, by describing how the performance of brickwork is affected by laying bricks with frogs up and down, provides an understanding of the reasons behind an architect's or engineer's instructions. It begins with some background information and concludes with comments on training practice.

WHY SOME BRICKS CONTAIN FROGS

For hundreds of years bricks made by hand and moulded by machine from soft clays have contained frogs, formed by inserts in the moulds, primarily to facilitate filling the moulds and demoulding the 'green' bricks (fig 3.25).

Figure 3.26.

Today, a large number of bricks are made from relatively dry granular clay by the semi-dry pressing process, using very great pressure. The process is facilitated by an insert in the mould which forms the frog (fig 3.26).

Bricks may have frogs in one or both bed faces, the latter generally being referred to as a double-frogged brick, in which case one frog is usually larger than the other.

The shape, size and number of frogs, if any, is mainly dependent on the characteristics of the clay.

FROGS UP OR DOWN? – THE EFFECT ON BRICKWORK PERFORMANCE

In considering the effect on the performance of brickwork the critical factor is not whether bricks are laid with the frogs up or down but whether the frogs, or the larger, in the case of double frogs, are filled solidly or not.

It is possible, although time consuming, to lay bricks with frogs down and filled (fig 3.27), but it is only practicable with limited numbers. For instance, if handed cant bricks are not available for either side of an opening it is possible to invert those on one side (fig 3.28).

Figure 3.25.

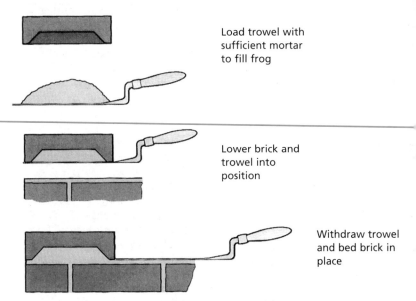

Load trowel with sufficient mortar to fill frog

Lower brick and trowel into position

Withdraw trowel and bed brick in place

Figure 3.27. Laying frog down and filled in exceptional circumstances.

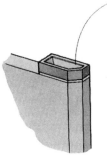

Figure 3.28. *Cant brick inverted if 'handed' versions not available.*

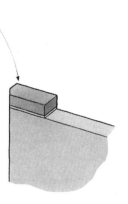

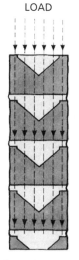

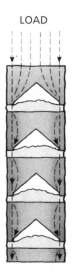

LOAD LOAD

Load evenly distributed throughout wall by mortar in frogs

Load concentrated at edges by lack of mortar in frogs

Figure 3.29. *Frog up is stronger.*

But this may not be visually acceptable with some textured bricks. *(see Section 2.9 'Bricks of special shapes and sizes', fig 2.86 and text).* If it proves difficult to 'wipe' mortar into the frog the alternative method described above should be used.

A number of relevant functions are described below together with the effect of laying the bricks with frogs up and down.

Strength and stability

Compressive strength tests on individual bricks generally require the frogs to be filled unless stated otherwise. Engineers use these results when calculating the loads that brickwork can support. Bricks without filled frogs will fail at much lower loads, as will the brickwork *(fig 3.29)*. One manufacturer states the compressive strength of his bricks to be 21 N/mm^2 frog up and 7 N/mm^2 frog down.

If bricks are not laid frog up and filled as instructed, the brickwork will not have the strength intended by the engineer and may crack, spall or even collapse. Examples of brickwork that must be designed and built to support heavy loads are walls, columns and piers under the bearings of concrete or steel beams and possibly lintels *(figs 3.30 & 3.31)*. Padstones are usually built-in to prevent the edges of bricks spalling when subjected to concentrated loads from wide spanning and deflecting beams.

Walls supporting concrete floors, especially in multi-storey buildings, will almost certainly be required to be laid frog (or larger frog) up.

Bricks in most housing and non-load-bearing walls in framed buildings are likely to have a

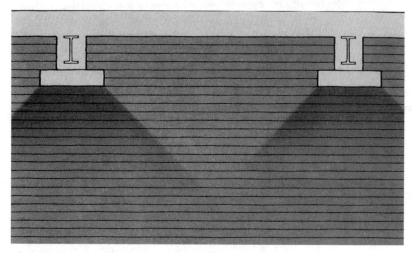

Figure 3.30. *High stresses in brickwork under beam bearings.*

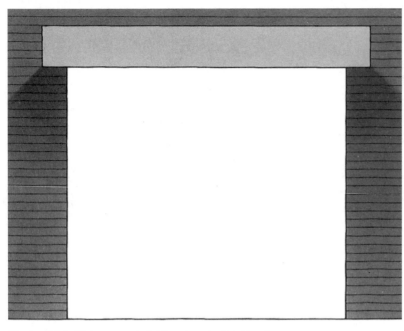

Figure 3.31. *High stresses in brickwork under lintel bearings.*

compressive strength much greater than that required, even when laid frog down, but instructions may still be given to lay bricks with frog up for other reasons, such as improved sound insulation.

Sound insulation
Effective sound insulation between adjacent buildings and rooms within a building depends largely on the density of the wall between them. Heavier walls contribute to better sound insulation and this may be a good reason for building them with the frogs up and filled solidly with mortar.

Fixings to brick walls
Smaller fixings such as those secured by plastic plugs or rubber bushes are generally satisfactory whether the frogs are filled or not. Larger expansive bolts must be used with care so as not to disrupt the brickwork whether or not the frogs are filled.

Other considerations
It is sometimes argued that the air pockets in unfilled frogs improve the thermal insulation of a wall. In practice the difference is negligible and may be ignored.

A BRS Digest written in 1954 stated that there was no significant difference in the rain resistance of a solid brick wall whether the bricks were laid frog up or down. The Brick Development Association is not aware of any subsequent evidence which contradicts these findings.

Some manufacturers of handmade bricks recommend laying them frog up, so that the

surface creases 'smile', helping to shed water and improve the durability of the bricks.

TRAINING PRACTICES
Some tutors insist that trainees, when laying bricks frog-up, fill the frogs prior to spreading the mortar bed because at first they have difficulty in putting down enough mortar and keeping to gauge. Other tutors prefer to teach from the beginning that trainees put down enough mortar to fill the frog and lay a full bed in one action as they must do in practice.

AUTHORITATIVE GUIDANCE
Bricklayers and supervisors should be aware of authoritative guidance available on matters related to their craft. In the event of a dispute over workmanship such guidance is likely to be taken into account. Some relevant guidance is quoted below.

Bricklayers bear the responsibility for building brickwork to achieve the designers' aims. Failure to do so will result in poor performance and the possibility of expensive remedial work.

KEY POINTS

- Always lay bricks frog up if so instructed.
- If no Instructions are given – ask.

- If there is no one to give instructions – lay frog up, especially under heavy loads.

'Laying bricks with frogs. Lay single frog bricks with frog uppermost and bricks with a double frog with deeper frog uppermost. Fill all frogs with mortar if specified. Lay bricks with frogs down only if permission is given. COMMENTARY. *Brick walls built with frogs down and unfilled are weaker and less resistant to sound transmission. Advice should be sought as to whether unfilled frogs are acceptable.* Clause 3.1.3.3, BS 8000:Part 3:1989 'Workmanship on building sites'.

'Bricks with frogs. Bricks should normally be laid on a full bed of mortar with the frog or larger frog uppermost which should be filled as the work proceeds'. Clause 8.2, BS 5628:Part 1:1978 'Structural use of unreinforced masonry'.

'Bricklaying. Unless otherwise specified, frogged bricks should be laid frog up and the frog filled with mortar. The position and filling of the frogs are important, as both can affect the strength and sound insulation of the wall'. Clause 32.7, BS 5628:Part 3:1985 'Use of masonry. Materials and components, design and workmanship'.

'Filling of joints and frogs. Single-frogged bricks shall be laid frog uppermost and double-frogged bricks shall be laid with the deeper frog uppermost. All frogs shall be filled with mortar'. Clause 2.4.3, SP 56 1988 'Model specification for clay and calcium silicate structural brickwork'. Published by British Ceramic Research Ltd.

'Frog up or frog down? Bricks must be laid frog up with all joints filled when maximum strength or weight is required for the brickwork. When neither is the prime requirement, the bricks may be laid frog down. If separating walls are required to meet Building Regulations for sound insulation, bricks should be laid frog up to give maximum weight'. 'Oxford clay fletton range technical information on brickwork'. London Brick 1987.

3.5   MANHOLES AND INSPECTION CHAMBERS

Brickwork in manholes and inspection chambers will be out of sight after backfilling, but must be built with care as faults can make effective maintenance difficult and are usually expensive to repair. If chambers leak excessively during water pressure tests they will have to be repaired or even rebuilt.

This section provides a basic guide only. Practice may vary in different parts of the country being usually determined by local building control officers.

FUNCTIONS AND REQUIREMENTS

Covered inspection chambers and manholes are built over drains and sewers to allow easy access for inspection, testing, maintenance and cleaning. A manhole is an inspection chamber within which a person can work.

All chambers should be:

- big enough to allow for the connection of branch drains to a main drain
- virtually watertight
- able to carry all expected vertical and lateral loads

Recommendations for design, specification and construction are given in British Standard Codes of Practice 8301[1], 8000[2] and 5628[3].

Note: The depth of a chamber is taken from the top surface of the cover to the invert level of the downstream end of the channel *(fig 3.34)*.

MINIMUM DIMENSIONS OF CHAMBERS

Based on the recommendations of BS 8301[1] table 8.

Chambers without branches *(fig 3.32)*.

The recommended minimum internal dimensions are:

- 450 mm long by 450 mm wide, if not more than 1 m deep *(fig 3.32a)*
- 1200 mm long by 750 mm wide, if more than 1 m deep *(fig 3.32b)*

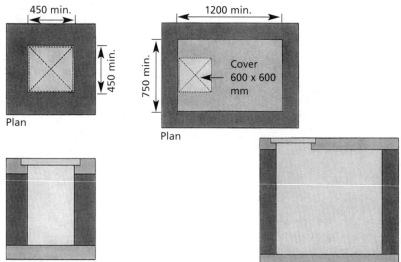

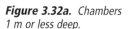

450 min.

450 min.

Plan

Vertical section

Figure 3.32a. *Chambers 1 m or less deep.*

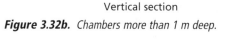

1200 min.

750 min.

Cover 600 x 600 mm

Plan

Vertical section

Figure 3.32b. *Chambers more than 1 m deep.*

Figure 3.32. *Minimum dimensions for chambers without branches.*

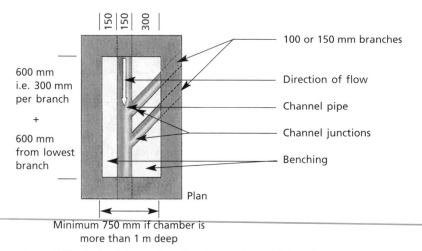

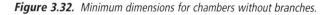

150 | 150 | 300

600 mm i.e. 300 mm per branch

+

600 mm from lowest branch

Plan

100 or 150 mm branches

Direction of flow

Channel pipe

Channel junctions

Benching

Minimum 750 mm if chamber is more than 1 m deep

Figure 3.33. *Basic minimum dimensions for chambers with branches.*

Chambers with branches

(fig 3.33)
Length. A minimum of 300 mm is allowed for each nominal 100 or 150 mm branch at the side having most connections. An additional allowance, usually of 600 mm, is made at the downstream end from the lowest branch, to allow for a connection to the channel junction and for rodding.
Width. The minimum width should be 150 mm, or the main nominal pipe size, if it is larger, plus 150 mm for a side with no branches plus 300 mm for a side with branches, **BUT**, note that

the total width must be not less than 750 mm if the chamber is more than 1 m deep.

Minimum nominal dimensions of chamber covers
* 450 mm by 450 mm for chambers not more than 1 m deep and
* 600 mm by 600 mm for chambers more than 1 m deep *(fig 3.32)*.

BASE SLABS
Base slabs must be designed and built to support the brickwork and vehicular traffic. Slabs will normally be 150 mm thick, although 100 mm may be permitted in domestic work where there is no risk of vehicular traffic loads.

Bases need not project beyond the external face of brickwork. The concrete mix will normally be grade C20P or a standard mix ST4 to BS 5328[4]. If you are site-batching by volume, ask for guidance on proportions of cement and aggregates to be used *(fig 3.34)*.

WALLS
Bricks The masonry Code of Practice[3] permits the use of any clay brick to BS 3921[5]. But some authorities require the use of Class B clay Engineering bricks. For drains carrying chemically aggressive fluids, class A or B clay Engineering bricks may be specified. Some authorities will not permit textured bricks to face into the inspection chamber.

Generally, calcium silicate bricks to BS 187:1978[6] are permitted but some types are not suitable for use with foul drainage. If in

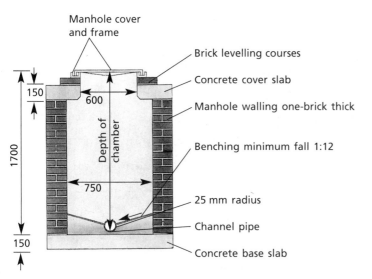

Figure 3.34. *Section through typical manhole looking up stream.*

Labels: Manhole cover and frame; Brick levelling courses; Concrete cover slab; Manhole walling one-brick thick; Benching minimum fall 1:12; 25 mm radius; Channel pipe; Concrete base slab; 150; 600; Depth of chamber; 1700; 750; 150

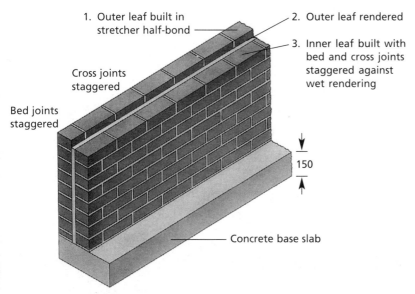

Figure 3.35. *Water or manhole bond.*

Labels: 1. Outer leaf built in stretcher half-bond; 2. Outer leaf rendered; 3. Inner leaf built with bed and cross joints staggered against wet rendering; Cross joints staggered; Bed joints staggered; Concrete base slab; 150

doubt ask the supplier or manufacturer *(see Section 6.4 'Durability of Brickwork' table 6.5 masonry condition 'L')*

Mortar Mortar should be designation (i) i.e. 1:¼:3 cement:lime:sand. Ensure that mortar is accurately batched and properly mixed otherwise it may suffer sulfate attack. Some

specifiers may require you to use sulfate-resisting cement with FN and MN quality clay bricks or if the ground contains sulfates. *(see Section 6.4 'Durability of Brickwork' table 6.5, masonry condition 'L')*

Brickwork Walls will normally be one brick thick to resist ground and water pressures. Half-brick

thick walls **may** be permitted if the brickwork is above the water table, the chamber is not more than 900 mm deep and will not be subject to vehicular traffic.

One-brick walls are normally built in English bond, but if the ground water pressure is high, 'water or manhole' bond may be specified to give increased resistance to water penetration *(fig 3.35)*.

BENCHING

Benching is the raised concrete surface between channels and walls *(fig 3.34)*. It must slope sufficiently to direct overflowing sewage back into the channels but be flat enough to provide a safe foothold when rodding inside a manhole. The absolute minimum slope is 1:12, but a little steeper is preferable.

COVER SLABS

Concrete cover slabs used to roof manholes are generally 150 mm thick, and must be designed and built to carry all loads including traffic. They should never be less than 100 mm thick in order to provide sufficient cover for reinforcement *(fig 3.34)*.

They are often set well below finished ground level to allow access covers and frames to be bedded to a road camber or to allow future alterations of ground level without disturbing the slab.

METHOD OF CONSTRUCTION

Support of trenches

People die in quite shallow trenches that collapse without warning. Unless the sides are

battered to the angle of repose,
ensure that suitable support is
provided for all trenches 1.2 m
or more deep and for shallower
trenches in which you or others
will have to work substantially
below ground level. The angle of
repose depends on the type and
condition of the ground, so if in
doubt seek advice from a
supervisor.

Part 14 of the code of practice
for 'Workmanship on building
sites'[2] makes recommendations.

Loading-out work area

Because the face side is internal,
it is easier to work from the
inside of manholes, standing
astride the channel. Load out
bricks around the four sides with
low stacks because the trench
sides may be unstable. One
mortar spot board is normally
adequate.

Levelling from datum *(fig 3.36)*

Level across from the datum peg
or paving. Gauge down to the
concrete base, marking on the
gauge rod:

* Thickness of cover and frame
  plus bedding mortar
* Levelling courses
* Cover slab or start of
  corbelling
* Remaining courses of
  brickwork

Setting out

* Spread a thin skim of mortar
  on the concrete base *(fig
  3.37)*.
* Mark position of internal wall
  face in the mortar, adjusting
  the dimensions to avoid
  broken bond.
* Set out the first course dry,
  normally in English bond, to

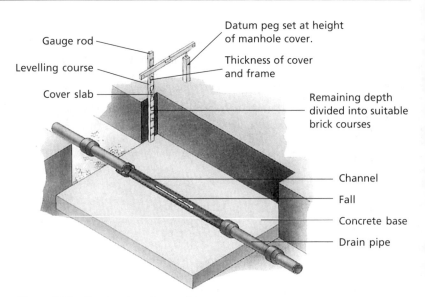

Figure 3.36. *Gauging down from a datum peg.*

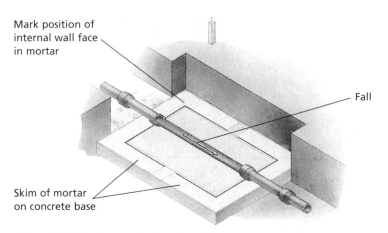

Figure 3.37. *Setting out the internal wall face.*

minimise the number of cut
bricks, especially if very hard
bricks are to be used.
* You may be told on site to
  build toothed quoins *(fig
  3.38)*, in order to avoid
  cutting queen closers from
  very hard bricks, particularly if
  a bench saw is not available
  at that time. This method
  cannot be used if the
  chamber is to be rendered
  externally. Furthermore, great

care should be taken to avoid
dislodging the projecting and
therefore vulnerable corner
bricks either before or during
backfilling.

Building the walls, pipework, benching and step irons *(fig 3.39)*

* Bed the first course of bricks
  to establish the bond.
* Solidly bed connecting pipes
  and build-in as the brickwork
  proceeds.

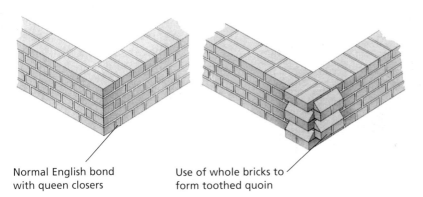

Normal English bond
with queen closers

Use of whole bricks to
form toothed quoin

Figure 3.38. *Alternative method of forming a quoin, compared with the normal method.*

- Lay bricks with single frogs or the larger of double frogs uppermost.
- Many specifiers will require nominal 10 mm bed and cross joints as recommended for brickwork in general in BS 5628[3] or specifically, for chambers, by the Water Research Centre[7]. Some specifiers may follow the recommendations in BS 8301 that joints should be not more than 6 mm and not less than 4 mm thick. It should be noted that the same Code of Practice recommends that bricks used in chambers,

besides being Class B to BS 3921[5], should additionally be subject to the limitation of twist and bow to ±2 mm.
- Take particular care to fill all joints solidly across the wall width. If chambers, when tested, lose more water than is permitted, the leaks are most often through badly filled joints and so the bricklayers are to blame.
- Although 'full and flush joints' are often specified, BDA recommends that joints are 'tooled' on both sides of the brickwork in order to maximise water resistance,

particularly if the ground water level might rise above the bottom of the chamber.
- After a couple of courses, stand on a plywood off-cut resting on dry bricks. This will be easier on the feet and will protect the channels and pipes from accidental damage.
- Raise about 6 courses, keeping internal corners plumb. Check for square **inside** the chamber as soon as the brickwork rises above the pipes.
- Form the benching at this point as the higher the brickwork the more difficult it becomes. Rough-out shape of benching in solidly bedded brickwork or using a stiff concrete mix, as described under 'Base slabs'.
- Rise vertically from the edge of the channel forming a 25 mm radius and slope up to the walls at not less than 1:12, but remember that it must be possible for others in the future to stand on the benching comfortably and safely whilst working in the chamber *(fig 3.39)*. Usually the final 25 mm of benching is finished with a 1:3 cement:sand render applied and trowelled smooth while the concrete is still green.
- Replace plywood when benching is hard, to protect it and the channels.
- Pipes of 300 mm or more in diameter should be protected either by a one-brick relieving arch turned over the pipe for the full thickness of the wall or a reinforced concrete lintel *(fig 3.40)*.
- Build in step irons[8] every fourth course into the wall

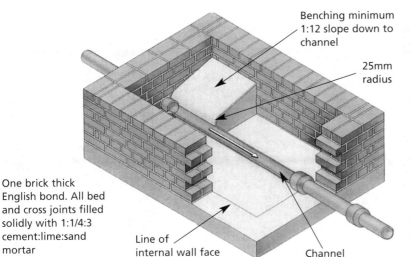

Benching minimum
1:12 slope down to
channel

25mm
radius

One brick thick
English bond. All bed
and cross joints filled
solidly with 1:1/4:3
cement:lime:sand
mortar

Line of
internal wall face

Channel

Figure 3.39. *Brickwork and benching.*

Brick-on-edge relieving arch

Concrete lintel

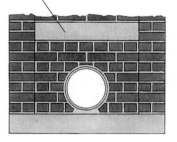

NOTE: For clarity no concrete benching is shown

Figure 3.40. *Methods of protecting large pipes of 300 mm and more.*

beneath the opening in the cover slab as illustrated *(figs 3.41 & 3.42)*.

- Continue brickwork plumb to the level of the underside of the concrete cover slab, which is usually 300–400 mm below ground level. Shutter and cast in situ concrete cover slab with reinforcement as detailed and specified.
- Alternatively, corbel out in headers in order to reduce the plan size of the manhole to the size of the cover.

Corbelling should not exceed 30 mm in each course.

- The access cover is positioned at the downstream end to facilitate rodding upstream *(fig 3.41)*.

Bedding cover frames *(fig 3.42)*

- Raise levelling courses to bring the frame and cover up to the required level.
- Solidly bed the frame in 1:3 cement:sand mortar with the cover in place to prevent the frame twisting. Take care to get the top surface exactly level

with adjacent paving, datum peg or to suit the road camber.

- Remove cover and neatly point up around inside of frame.
- Replace cover and form fillet around frame to protect exposed edge.

References
(1) BS 8301:1985 'Code of practice for building drainage'.
(2) BS 8000:Part 14:1989 'Workmanship on building sites. Codes of practice for below ground drainage' and Part 3:1989 'Codes of practice for masonry'.
(3) BS 5628:Part 3:1985 'Code of practice for use of masonry, materials and components, design and workmanship'.
(4) BS 5328:1990 'Concrete: methods for specifying concrete'.
(5) BS 3921:1985 'Specification for clay bricks'.
(6) BS 187:1978 'Specification for calcium silicate (sandlime and flintlime) bricks'.
(7) 'Sewers for adoption – design and construction guide for developers'. Water Research Centre (1989) plc for the Water Authorities Association.
(8) BS 1247:1975 'Manhole step irons'.

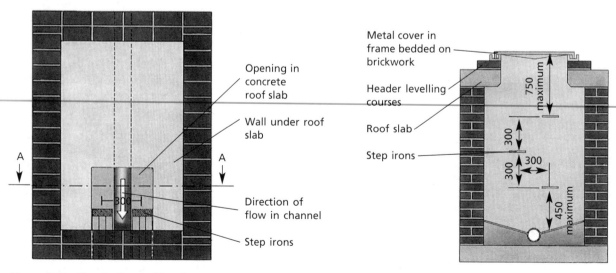

Opening in concrete roof slab

Wall under roof slab

Direction of flow in channel

Step irons

Figure 3.41. *Plan showing position of cover opening and step iron.*

Metal cover in frame bedded on brickwork

Header levelling courses

Roof slab

Step irons

Figure 3.42. *Section AA through manhole looking downstream showing positions of step irons.*

KEY POINTS

- Ensure trench sides are safe – ask supervisor if in doubt.
- Gauge down from datum peg marking position of access cover and frame, levelling courses and cover slab on a guage rod.
- Mark internal size of chamber on base slab.
- Load out with low stacks of bricks around excavation.

- Set out brickwork bond to minimise cutting – especially if hard bricks are specified.
- Fill all bed and cross joints solidly.
- 'Tool' joints internally and externally.
- Check internal corners for plumb and square after raising about six courses.

- Slope benching down to channel at least 1:12 but not too steep.
- Replace plywood working panel to protect benching and channels.
- Position correct access cover and any step irons at downstream end.
- Bed cover frame solidly, point up internally and form fillet externally.

4 ACCESSORIES

The variety of components already introduced in the foregoing sections as typical examples are dealt with here in greater detail. The section also includes articles on aspects of construction that are frequently encountered in modern construction and involve more complex construction detail than traditional work – building brickwork on metal support systems as cladding for framed structures, reinforcement in masonry structures. The incorporation of movement joints into brickwork is also detailed; this is an aspect of modern work that is frequently misunderstood and incorrectly assembled.

4.1　MORTARS

Mortar is a mixture of materials used in bedding, jointing and pointing bricks and blocks in masonry walling *(fig 4.1)*.

Mortar consists of sand; a binder, usually ordinary Portland cement (OPC); water and frequently a plasticiser. Hydraulic lime, commonly used as a binder at one time, is rarely used today in normal building work, but it is sometimes used for the restoration and repair of old brickwork.

When hydraulic lime mortar is specified its composition requires particular care. Allowance for longer setting times should also be made.

THE IMPORTANCE OF WORKMANSHIP

Good designers specify the right mortar for strong, durable, rain resistant, crack free and attractive brickwork.

BUT

They rely on the care and skill of well-informed bricklaying teams to achieve these ends.

WHAT IS REQUIRED OF MORTAR?

Workability

Bricklayers require a 'fatty' mortar which hangs on the trowel without being sticky, spreads easily and stiffens neither too quickly nor too slowly *(fig 4.2)*.

BUT

In achieving workability, bricklayers must also produce a mortar to meet the life-long requirements of brickwork.

Life-long requirements include:
- adequate compressive strength
- adequate bond strength between mortar and bricks
- durability – i.e. resistance to frost and chemical attack
- joints sealed against wind-driven rain
- a pleasing appearance.

Figure 4.1. *Spreading mortar.*

Figure 4.2. *A 'workable' mortar.*

The ability of mortars to meet these requirements depends on:

- THE MATERIALS SPECIFIED – by the designers
and
- THE WORKMANSHIP – of the bricklaying team, including the protection of materials and brickwork against bad weather conditions.

MATERIALS
As delivered for on-site batching.

Architects, engineers or surveyors are responsible for specifying the right type and quality of materials. The bricklaying team is responsible for storing, protecting, proportioning, mixing and using them with care.

Sand
Sands for bricklaying mortars are normally dug from a pit. Sea sands contain salts which adversely affect the quality of the mortar. They should not be used unless they have been washed and supplied by a reputable firm.

Good mortar sands are well graded having fine, medium and coarser particles.

Poorly graded sands, with single size aggregates, contain a greater volume of air and require more binder to fill the spaces

and make the mortar workable *(fig 4.3)*. Mortar made from poorly graded sands will be weaker, unlikely to retain fine particles and more likely to shrink, leaving cracks for rain to penetrate.

- Use only the specified sand. Different sands may require different mix proportions.
- Obtain all the sand for the job from one source. Different sands can result in different mortar colours and patchy brickwork.
- Store and protect sand from rain and contamination by other building materials, mud, vehicles and plant. Dirty sands produce weaker, less durable and discoloured mortars.

Cement
The most commonly used binder in bricklaying mortars is ordinary Portland cement (OPC). 'Masonry cement' is frequently used but remember it contains only 75% OPC, the remainder being an inert filler, which has no binding capability.

- Use only the specified cement.
- Obtain all the cement for the job from one source. Different cements can result in different mortar colours and patchy brickwork.

- Never use high alumina cement.
- Never substitute masonry cement for Portland cement. Different mix proportions will have to be used and there will be a change of colour.
- Store bags off the ground and protect from rain *(fig 4.4)*.
- Do not use cement which has been exposed and contains 'lumps'. It will produce weaker and less durable mortars.

Hydrated limes
In the 19th and early 20th centuries, *hydraulic* limes were used as the only binding agent in mortars. Even if these were still generally available they would not be practical for the majority of modern building as they harden slowly rather than set quickly as cement mortars do.

Today *hydrated*, non-hydraulic limes in powder form are often added to Portland cement-based mortar to improve workability. Being water retentive, lime also improves the bond with the bricks and therefore the brickwork's tensile strength and rain resistance.

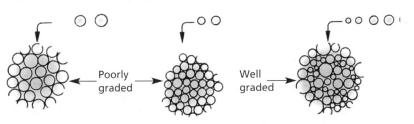

Figure 4.3. *Simplified diagrams of sand grading.*

Figure 4.4. *Protect cement bags.*

Figure 4.5. *Detergents make poor mortar.*

Figure 4.6. *Container in use with top cover removed.*

Plasticisers

Cement, sand and water alone often produce harsh mortars which are difficult to use. Their workability can be improved by adding lime or proprietary plasticisers which entrain (trap) minute bubbles of air in the mortar.

- Use only proprietary plasticisers as specified or permitted.
- Follow the manufacturers' recommendations regarding quantities and mixing times precisely.
- Never use domestic or commercial detergents as they may contain harmful chemicals. Unlike proprietary plasticisers, they generate uncontrolled quantities of large air bubbles producing weaker, less durable mortars *(fig 4.5)*.

Pigments

Pigments are added to produce 'coloured' mortars.

- It is virtually impossible to add accurate proportions on site and maintain colour consistency.
- Pigmented mortars are usually delivered to site as 'dry' ready-mixed lime:sand ('coarse stuff') or 'wet' ready-to-use retarded cement mortars.

WATER

- Water used for mixing should be clean enough to drink.

Ready-mixed lime:sand for mortars

- Properly proportioned and mixed lime:sand for mortars is delivered 'dry' in bulk to sites by specialist suppliers. The mixes may also include plasticisers and pigments. Ready mixed lime:sand is convenient to use but must be treated with care.
- Protect from contamination by other materials, mud and vehicles *(fig 4.6)*.
- Protect from extremes of weather. Fine particles of lime and pigments can be washed away by rain or blown away by drying winds causing variations in mortar colour.

Ready-to-use retarded mortars

These mortars are delivered 'wet' to site usually in covered containers of 0.25 or 1 m^3 capacity *(fig 4.7)*. They can be used without further additions or mixing for up to thirty six hours i.e. two working days. It is good practice to:

- Cover containers to minimise the effects of weather, prevent contamination by vandalism and accidents to children. Some containers are made so that they may be padlocked.
- Clean containers before refilling to avoid contamination of fresh mortar and renew plastic liners if provided.
- Avoid re-tempering the mortar once the initial set has begun. If in doubt consult the supplier.

Figure 4.7. Delivery of ready-to-use retarded mortars.

'General Use Mortars' which are mortars made with cement, lime and sand with air-entrainment.

- The General Use Mortars are a relatively new development in mortar technology. The combination of adding lime and air-entrainment to Portland cement-based mortars enhances their adhesion, workability and durability. It also makes them less susceptible to variations in sand grading and thereby simplifies the mix specification.

WORKMANSHIP[3]

Gauging
Accurate gauging (measurement) of the quantities of mortar ingredients before mixing is

Describing mortar mixes
Mortars are specified as:

- proportions of specific materials, e.g. 1:1:6 cement:lime:sand or
- by a 'designation' number, (i), (ii), (iii) or (iv). This permits

the use of one of three mortar types and these are to be made with one of four binder mixes. These are set out in table 4.1 which is based on information taken from the Masonry Code of Practice[1] with the addition of

TABLE 4.1	Mortar mixes				
Basic composition	Cement:lime:sand with air-entrainment		Cement:sand with air-entrainment		Cement:lime:sand
Binders	Ordinary Portland cement or sulfate-resisting Portland cement	Masonry cement with high lime content (1:1 OPC:lime)	Masonry cement with other than high lime content	Ordinary Portland cement or sulfate-resisting Portland cement	
Designation					
(i)					1 : 0–¼:3
(ii)	1 : ½ : 4½ + Air	1 : 3	1 : 2½–3½	1 : 3–4 + Air	1 : ½ : 4–4½
(iii)	1 : 1 : 5½ + Air	1 : 4½	1 : 4–5	1 : 5–6 + Air	1 : 1 : 5–6
(iv)			1 : 5½–6½	1 : 7–8 + Air	1 : 2 : 8–9

     – Mortar of high durability   – General use mortar of good durability

NOTES:
The types of mortars of any one designation are of approximately equal strength. The range of sand proportions is to allow for varying grades of sand. The second quantity e.g. 1:1:5–6 for designation (iii) is for a well-graded sand. Smaller proportions of sand (or large proportions of cement and lime) are necessary with less well-graded sands *(see fig 4.3).*
The proportions of hydrated limes may be increased by up to 50% to improve workability.
With the permission of the designer, plasticisers may be added to lime:sand mixes to improve their early frost resistance. Ready-mixed lime:sand mixes may contain such admixtures. This table is based on information given in Table 15 of BS 5628:Pt3:1985.

Figure 4.8.

essential if the required mortar strength, durability and consistency of colour is to be achieved.

Weigh batching is the most accurate method but is seldom used, except on large sites, for practical or economic reasons. Measurement by volume, if done with care, is generally adequate. *Site batching – all ingredients mixed on site*.

If hand mixing:
- Use accurate and consistent proportions of materials otherwise the mortar and brickwork may not be sufficiently strong or durable and the colour will certainly vary.
- The use of shovels to proportion the materials is totally unsatisfactory. A shovelful of damp sand has a greater volume than a shovelful of dry powdery cement *(fig 4.8)*.
- Mortars batched this way often contain too little cement and this is the cause of a large proportion of brickwork failures investigated by BDA and BRE.
- Proportioning is best done by using bottomless steel or timber gauge boxes on a flat clean mixing surface or alternatively 10 litre steel buckets. A separate container should be used for cement.
- Mix all the ingredients dry by 'turning' the heap three times.
- Hollow the centre and add the water, gradually mixing it with the mortar.
- Finally 'turn' twice for a complete mix and to improve workability.
- If using hydrated lime the best results are obtained by mixing it with sand before adding water. Allow the resulting 'coarse stuff' to stand for at least sixteen hours (overnight). Gauge with cement immediately before use. 'Coarse stuff' need not be used immediately but should be protected from drying out.

If machine mixing:
- The first two points, under 'hand mixing' above, concerning batching, and the last point on hydrated limes apply equally to machine mixing.
- Clean the mixer thoroughly after use especially if pigments have been used.

- Use the correct amounts of water. Too much will produce weak mortar and may lighten the colour. An estimation of the amount of water required may be determined from table 4.2 which is an extract from the Masonry Code of Practice.[2]

Various experienced bodies advocate slightly different mixing sequences. However, there is agreement that it is unsatisfactory to add the cement, lime and sand to the mixer and mix them before adding the water. Tests have shown that much of the cement is left clinging to the side of the mixer drum, leading to cement-lean mortars.

Two advocated methods are noted below:

The first is based on the recommendations in the British Standard Code of Practice for 'Workmanship on building sites'[3]

- Load three-quarters of the required water and sand or premixed lime:sand. While mixing add the cement, or cement:lime, gradually and allow to mix in.

Table 4.2. Estimation of quantity of water required.

DESIGNATION	MEAN WATER DEMAND – LITRES/50kg CEMENT		
	cement: lime:sand	masonry cement:sand	cement:sand & plasticiser
(i)	40	–	–
(ii)	50	35	40
(iii)	70	45	50
(iv)	100	55	60

NOTE: The information in this table has been extracted from Table 18 of BS 5628:Pt3:1985.

- Finally, add the rest of the sand or premixed lime:sand and water necessary to achieve workability.

There is a view that unless the cement is added very carefully and gradually it will 'ball' and not be evenly dispersed, leading in the longer term to staining from the mortar. To avoid this possibility others advocate the second method:

- Add three-quarters of the required water and gradually add the required cement slowly to ensure a thin paste free from lumps.
- Add the remaining materials and water.

Whichever method is used:

- Mix each batch for a consistent length of time. Three to five minutes, after all the constituents have been added, is sufficient. A short period will produce a non-uniform mix having poor workability. A long period will produce a weaker mortar having a poorer bond.
- Do not load the mixer to more than its rated capacity.

Admixtures
- Use admixtures only with the permission of the designer.
- When using plasticisers or masonry cement (which contains plasticisers) do not add too much water at the start as these mortars become very fluid as air is entrained.
- Follow the manufacturers' instructions.
- Proposed admixtures should be tested in mortar on a test panel because some of them affect the working life and/or modify the colour of mortars. If either effects occur the architect or supervisor should be informed.

Site batching – using ready-mixed lime:sand
- Care in gauging the cement accurately is just as essential as when mixing all the ingredients on site.
- Add three-quarters of the required water and gradually add the required cement slowly to ensure a thin paste free from lumps.
- Add the coarse stuff and water to achieve workability.

BRICKLAYING
- Water may be added for workability if the mortar becomes dry but it should never be retempered once the initial set has begun. This is particularly important when using ready-to-use mortars.
- When required to iron the finished joints (bucket handle or weather struck), it is bad practice to iron all the joints in one operation at the end of the day. The mortar placed at the beginning of the day may be dry resulting in the ironing tool 'dragging' the face of the mortar. The recently completed joints, particularly if low absorbency bricks are used, will be fluid and ironing will bring the fine particles to the surface with the risk of lightening the surface and staining the adjacent brick faces. Every member of the team should use a similar technique otherwise the appearance will vary.
- Newly built brickwork must be protected until the mortar has set. Rain will wash out the fine particles of cement, lime and pigments. This will change the colour of the mortar, the apparent colour of the brickwork and will possibly cause permanent staining. Frost will permanently damage mortar which has not set.
- If frost is likely, protect newly built brickwork overnight with waterproof insulation. (*see Sections 1.2 'Protection of newly built brickwork'; 3.1 'Avoiding damage from extremes of temperature'*)

- Do not add 'anti-freeze' admixtures. Although effective in mass concrete they are not effective in brickwork because the large volume of bricks quickly depresses the temperature of the comparatively thin layers of mortar. Do not build masonry when air temperature is at or below 3°C and falling.[4]
- Do not lay mortar on frozen surfaces even if the air temperature rises above freezing.

Mortar is not merely 'muck' used to assemble bricks. It is vital in producing strong, durable and attractive brickwork. Good workmanship is vital to the production of good mortar.

References
(1) BS 5628:Part 3:1985 'Use of masonry' – table 15.
(2) *Ibid* – table 18.
(3) BS 8000:Part 3:1989, cl. 2.2.3.5 'Workmanship on building sites – code of practice for masonry'.
(4) *Ibid* – cl. 3.1.1.1.

KEY POINTS

- Use only specified materials.
- Gauge materials accurately.
- Neither undermix nor overmix.
- Protect materials.
- Protect brickwork until mortar has set.

4.2 TIES IN CAVITY WALLS

The bricklayer is responsible for building-in the type of ties which have been specified in the correct positions and in a competent workmanlike manner. Failure, through lack of skill, care or attention, may lead to damp penetration, distortion, cracking or in extreme cases collapse of the wall.

This section deals with ties commonly used in cavity wall construction. The more sophisticated techniques for supporting and restraining brickwork cladding to framed buildings will be described in another section.

THE PURPOSE OF WALL TIES
Ties allow the two slender leaves of a cavity wall to support each

other and produce a stronger wall than one in which the leaves stand independently *(fig 4.9)*. Two leaves acting together will be virtually as strong as a solid wall only if the bricklayer ties them together securely with an adequate number of suitable ties.

TYPES OF TIES
Ties are made in a number of different shapes and forms from galvanised steel, stainless steel, alloys or polypropylene. They are available in various lengths to suit different cavity widths *(fig 4.10)*.

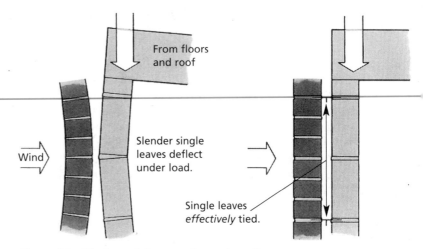

From floors and roof

Wind

Slender single leaves deflect under load.

Single leaves *effectively* tied.

Figure 4.9. Effective wall ties strengthen cavity walls.

Figure 4.10. *Types of ties.*

The ties will be specified by the architect or engineer for particular reasons and the bricklayer should not use any other tie without obtaining permission.

POSITION OF TIES

The designer will determine the distance between ties to give the required strength and possibly to suit the placing of thermal insulation batts or boards. It is not practicable for designers to specify the position of each tie and they must rely on the bricklayer to make common sense decisions in particular situations.

There are two useful rules:

- If in doubt ask.
- More is better than less.

The bricklayer may be required to space the ties, horizontally and vertically, in accordance with the requirements of the Building Regulations *or* the recommendations of the Masonry Code of Practice[1] which differ slightly. They are both illustrated here *(fig 4.11)*.

GOOD WORKMANSHIP IN PLACING TIES

- Bed all ties at least 50 mm into each leaf – for strength. Position any drip in the centre of the cavity and pointing downwards *(fig 4.12)*.
- Press ties down into the mortar bed – again for strength – do not place ties on bricks and then lay the mortar over them.
- Never push ties into a joint as they will not be effective in tying the two leaves together.
- Preferably incline ties (without deforming them) downwards to the *outer* leaf; never downwards to the *inner* leaf when they may provide a path for water and dampness across the cavity to the inner leaf and into the building *(fig 4.13)*.
- Maintain gauge and consistent thickness of bed joint.

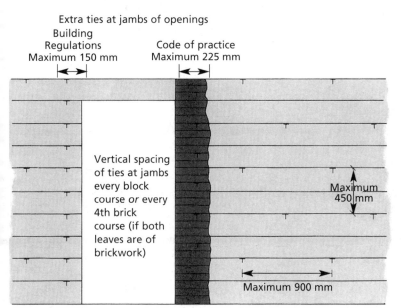

Extra ties at jambs of openings

Building Regulations Maximum 150 mm

Code of practice Maximum 225 mm

Vertical spacing of ties at jambs every block course *or* every 4th brick course (if both leaves are of brickwork)

Maximum 450 mm

Maximum 900 mm

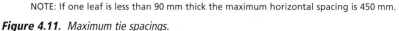

NOTE: If one leaf is less than 90 mm thick the maximum horizontal spacing is 450 mm.

Figure 4.11. *Maximum tie spacings.*

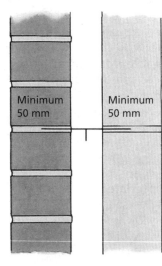

Minimum 50 mm

Minimum 50 mm

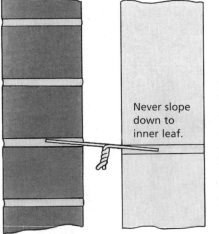

Never slope down to inner leaf.

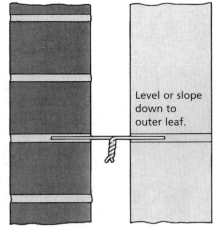

Level or slope down to outer leaf.

Figure 4.12. *Embedding ties.*

Figure 4.13. *Positioning ties.*

- Do not bend ties to suit coursing.
- Do not hammer ties, particularly if galvanised. Cracked galvanising will lead to rusting and a weakened wall. Discard ties with defective galvanising.
- Clear ties of mortar droppings which can conduct moisture across the cavity and into the inner leaf.

THERMAL INSULATION – CAVITY BATTS AND BOARDS

Insulating cavity batts **completely** fill the cavity; they are 455 mm high and are supported by **normal** wall ties placed in the bed joints 450 mm apart vertically.

Insulating cavity boards **partially** fill the cavity and are supported at 450 mm intervals vertically by **special** wall ties incorporating a device to keep the board against the inner leaf *(fig 4.14)*.

The problem of coursing the batts should be solved at ground level by carefully examining the drawings.

The first row of ties for the purpose of supporting insulation boards may need to be positioned below the damp-proof course. If in doubt the bricklayer should ask the brickwork supervisor.

Each batt or part batt must be supported by at least two ties and each board supported *and* restrained by at least two ties top and bottom. If the required minimum horizontal spacing of 900 mm, for structural purposes, is too wide for supporting the insulation the ties should be placed closer together.

Extra ties should be inserted where cut boards occur to ensure that they do not lean outwards and bridge the cavity. *(see Section 4.4 'Insulated cavity walls' for more information)*

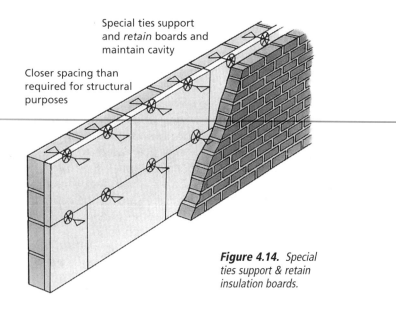

Special ties support and *retain* boards and maintain cavity

Closer spacing than required for structural purposes

Figure 4.14. *Special ties support & retain insulation boards.*

Figure 4.15. *Ties in a gable wall.*

And on a final note . . . The skill, care and attention exercised by the bricklayer in building-in ties is the last link in the chain producing strong, stable, durable, rain resistant walls.

Reference
(1) BS 5628:Part 3.

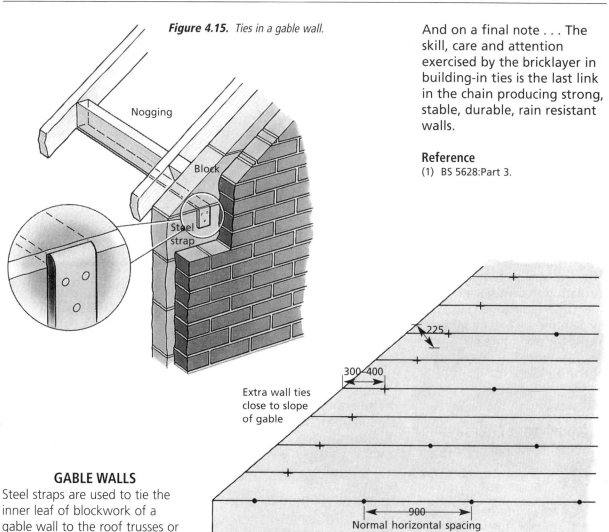

Figure 4.16. *Gable wall tied to roof structure.*

GABLE WALLS

Steel straps are used to tie the inner leaf of blockwork of a gable wall to the roof trusses or rafters to prevent the gable wall from being sucked out by the wind *(fig 4.15)*.

It is therefore important that the outer leaf of brickwork is securely tied to the blockwork *(fig 4.16)*.

The Building Research Station recommends that ties are positioned at least 300 mm vertically within 225 mm of the verge. In practical bricklaying terms this means that the end tie in every course of blocks should be within 400 mm of the roof line for roofs with a 35 degree pitch and within 300 mm for a 50 degree pitch.

KEY POINTS

- Use only the specified type of tie.
- Position ties strictly as instructed.
- Provide extra ties if needed to support insulation.
- Provide extra ties around openings.
- Provide extra ties at the tops of gable walls.
- Bed ties a minimum of 50 mm into each leaf.
- Position any drip in the centre of the cavity.
- Press ties into mortar bed. Do not push them into joints.
- Incline ties down towards the outer leaf.
- Do not damage galvanising by bending or hammering ties.
- If in doubt ask.

4.3   DAMP-PROOF COURSES

Bricklayers are responsible for building-in damp-proof courses, including cavity trays, to prevent the penetration of rainwater and ground moisture.

If this is not done with skill, care and attention, damp may cause timber to rot, plaster and decorations to deteriorate and the effectiveness of thermal insulation to be reduced.

Repairing damage caused by dampness is invariably very expensive and often distressing for the occupants of the building.

DAMP-PROOF COURSES – WHERE THEY ARE USED

DPCs are required in a number of places:

- At the base of external walls, not less than 150 mm above ground level.
- Similarly at the base of internal walls that are built off foundations rather than the ground floor slab.
- Vertically at jambs to openings in external cavity walls.
- Horizontally over openings in external cavity walls, where they are usually called cavity trays.
- Horizontally at window sill and door thresholds.
- Below copings and cappings to free-standing, retaining and parapet walls and chimney stacks.

DAMP-PROOF COURSES – THE MATERIALS USED

There are two main types of DPCs:

Flexible

Those most commonly used are of bitumen polymer, pitch polymer or polythene and are supplied in rolls in a range of widths including 110, 220 and 300 mm (fig 4.17).

Rolls of bitumen-based flexible DPCs should be stored on end to avoid distortion. They should be kept in a warm place, particularly in cold weather, to prevent cracking as they are unrolled.

Lead or copper DPCs, usually coated with bitumen to prevent corrosion and the possible staining of brickwork, are more costly and are used less often.

Rigid

Two courses of DPC bricks bedded in a cement-rich mortar are often used at the base of free-standing walls where they provide a better resistance to

overturning of the wall than do flexible DPCs. They may also be used at the base of external walls in buildings (fig 4.18a).

- Rigid DPCs of slate or tiles are seldom used today (fig 4.18b).
- Rigid DPCs are suitable to resist rising damp but not the downward flow of water.

DAMP-PROOF COURSES – BEDDING IN MORTAR

For flexible DPCs, a flat mortar bed should be laid to ensure that the DPC will be supported throughout its entire length and width. The mortar bed should not be furrowed (fig 4.19).

The mortar bed should be free from stones, pieces of brick or other hard objects and deep enough so that any projections or sharp edges will not perforate the DPC.

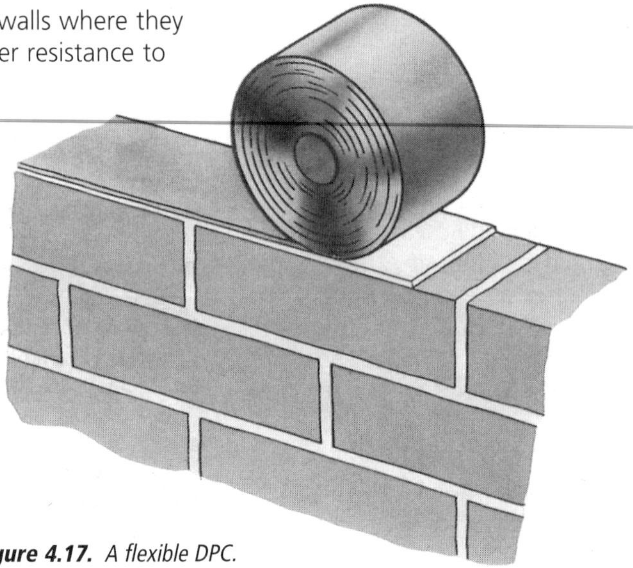

Figure 4.17. A flexible DPC.

Figure 4.18a. *Two courses of DPC bricks bedded in designation (i) mortar.*

Figure 4.18b. *Two courses of slates bonded and bedded in designation (i) mortar.*

The DPC roll should be placed on one end of the wall and carefully unrolled and pressed onto the mortar bed. This may be done by sliding a smooth brick or block along the DPC.

If the roll is not long enough or there is a change of direction, a new length of DPC should be laid to overlap by at least 100 mm. It is considered good practice to allow a lap equal to the width of the DPC. The overlap should be secured by a jointing compound or tape.

Care should be taken to use the correct DPC for the width of the wall. The DPC should extend through the full thickness of a solid wall and through each leaf of a cavity wall and should not be covered by pointing or rendering.

DAMP-PROOF COURSES IN CAVITY WALLS

DPCs should not project into the cavity where mortar droppings may build up and lead to moisture penetration *(fig 4.20)*.

VERTICAL DAMP-PROOF COURSES AT OPENINGS

When constructing windows and door jambs in a cavity wall, the bats that abut the window or door frames should be placed so that the smooth uncut surface is towards the DPC. This will reduce the risk of the cut edge damaging the DPC *(fig 4.21)*. When closing a cavity at the jambs, a mortar joint should be formed between the inner leaf and the DPC. To do this the bats used in the return will have to be buttered before they are placed in position making sure that the joint is full.

In many jobs, special cavity closer blocks may be supplied.

If a window or door frame is to be fixed after the opening is formed the vertical DPCs in the reveals should project beyond the edge of the brickwork by about 5 to 10 mm so as to make contact with the frame when it is in position. The DPC should project into the cavity by at least 25 mm.

Figure 4.19. *Bedding DPC.*

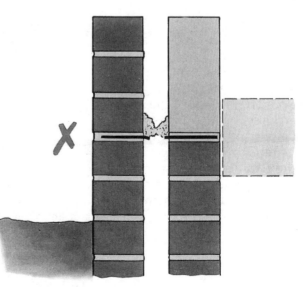

Figure 4.20. *Cavity obstruction.*

CAVITY TRAYS – WHERE THEY ARE USED

Cavity trays are used over window and door openings, at sills or any part of the construction which bridges the cavity. Their purpose is to prevent water from the outer leaf reaching the inner leaf, the cavity insulation, lintels or window and door frames *(fig 4.22)*.

The water collected by the tray will be drained by 'weep holes', usually in the form of open vertical cross-joints. Sometimes a special drainage tube or fibre filter is built into the joint.

If the trays cannot be formed in one length, any joints should overlap by at least 100 mm and be sealed to prevent water penetrating under the tray.

DPC trays should extend beyond the end of lintels and where specified be fitted with effective stop ends to prevent water draining into any cavity insulation which may be installed.

At jambs to openings trays should lap over vertical DPCs which in turn should overlap any trays below, such as those at sill level.

And on a final note . . .
No matter how well the architect has designed the details and chosen the materials, the bricklayer has the final responsibility for building brick walls which will keep out the rain and damp.

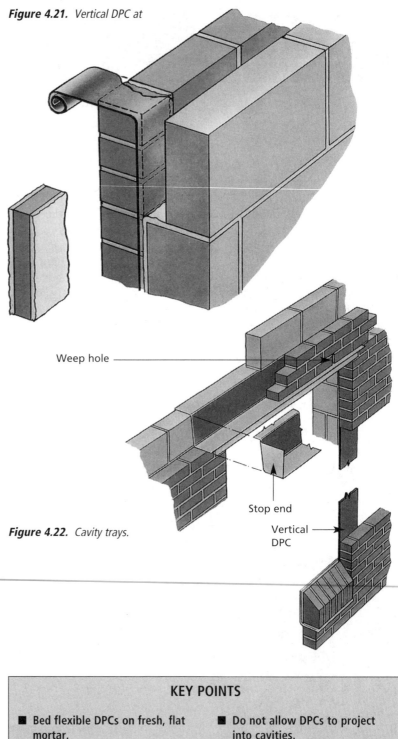

Figure 4.21. Vertical DPC at

Figure 4.22. Cavity trays.

Weep hole

Stop end

Vertical DPC

KEY POINTS

- Bed flexible DPCs on fresh, flat mortar.
- Any overlaps to be a minimum of 100 mm.
- Do not cover edge of DPCs with mortar or render.
- Do not allow DPCs to project into cavities.
- Extend DPC trays beyond end of lintels.

4.4 INSULATED CAVITY WALLS

The knowledge, understanding and care required to build an insulated cavity wall demonstrate the modern bricklayer's need for more than the basic bricklaying skills.

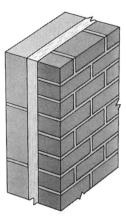

Figure 4.23a.
Traditional solid wall.

Figure 4.23b. *Traditional cavity wall.*

Figure 4.23c. *Insulated cavity wall.*

WHY CAVITY INSULATION?

The most common form of external wall used in recent years had a facing brick outer leaf tied to a 'light weight' concrete block inner leaf with an air space or cavity between. 'Cavity walls' gradually replaced solid walls during the first half of this century as experience showed that when properly designed and built they were more resistant to rain penetration than solid walls *(fig 4.23a & b)*.

Uninsulated cavity walls fall short of current insulation requirements which are most conveniently and effectively met by building insulating materials into the cavity *(fig 4.23c)*.

Well-designed insulated cavity walls will meet the high insulation standards and remain rain resistant only if carefully built, preferably by bricklayers who understand the special requirements.

CAVITY WALLS – GENERAL

Success begins with good basic construction techniques. These are dealt with more throroughly in sections 3.3 'External cavity walls'; 4.2 'Ties in cavity walls'; 4.3 'Damp-proof courses'.

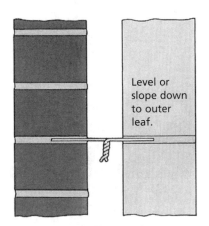

Level or slope down to outer leaf.

Figure 4.24a.

Some of the points are summarised here.

- Wall ties should be level or slope slightly down towards the outer leaf with the drip positioned in the centre of the cavity and pointing downwards *(fig 4.24a)*.
- Use only those ties specified, as they may be specially designed for use with insulating materials *(fig 4.24b)*.

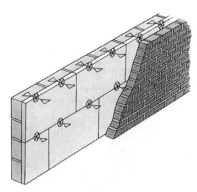

Figure 4.24b.

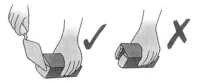

Figure 4.24c.

- Particular care must be taken to keep the cavities clear of mortar droppings when cavity insulation is to be used.
- The cross joints should be solidly filled with mortar to minimise rain penetration *(fig 4.24c)*.

- Provide DPC trays, with stop ends, over insulation that is not built right to the top of gable end walls *(fig 4.24d)*.
- Provide DPC trays over lintels etc., and carefully build-in stop ends to prevent water running off the end of the trays into the insulation *(fig 4.24e)*.
- Form weep holes to drain water effectively from the trays.

Figure 4.24d.

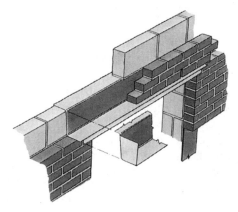

Figure 4.24e.

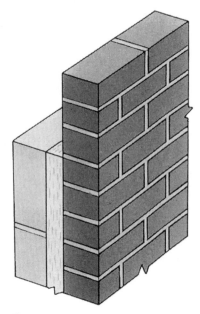

Figure 4.25a. *Full-fill cavity batts.*

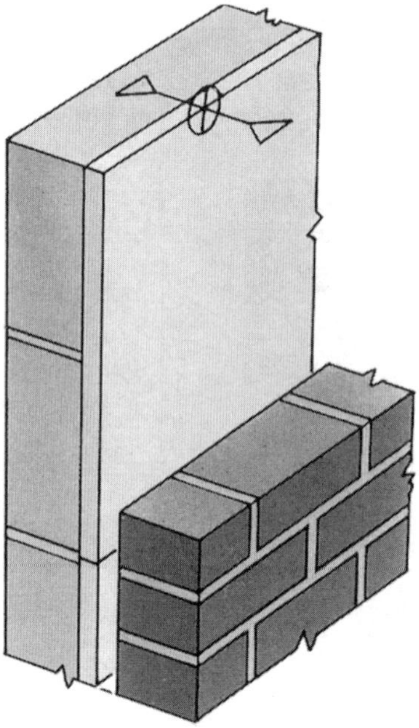

Figure 4.25b. *Partial-fill cavity boards.*

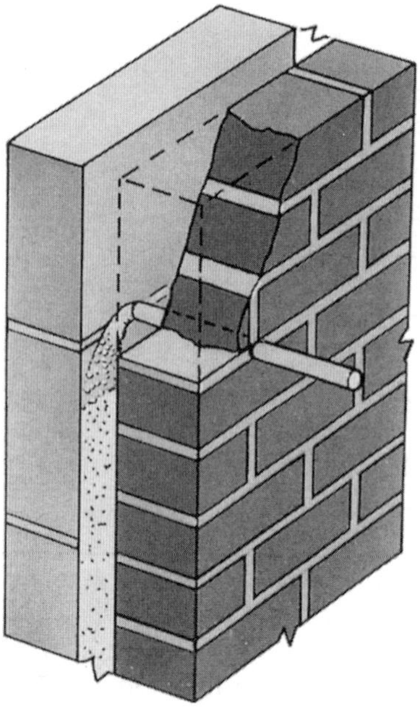

Figure 4.25c. *Injection of insulation.*

TYPES OF INSULATION

There are three methods of insulating cavities *(fig 4.25)*:
(a) Building-in full-fill cavity batts.
(b) Building-in partial-fill cavity boards.
(c) Injecting insulation after construction.

The first two methods require the bricklayers to cut, fit and fix the insulation as they build the cavity wall.

The third method requires the bricklayer to build the cavity wall with care, paying particular attention to any special requirements of the installers.

The recommended techniques for each method are described below.

THINGS TO BE DONE BEFORE BUILDING BEGINS

- Check that there is provision for storing and protecting the insulation when delivered.
- Boards for partial fill should be stored flat, never on bearers. Twisted boards will be difficult to fit closely to the inner leaf.

- Check that there is provision for protecting partly built work and any insulating materials that have been unwrapped ready for use, from rain and snow.
- When the insulation is delivered check that it is the correct type, height and thickness.

1. BUILDING-IN FULL-FILL CAVITY BATTS

Materials
Batts are soft and flexible, 455 mm high by 1200 mm long and of various thicknesses. They are made from layers of mineral fibres treated with a water repellant. Water will not penetrate through the batts but drain down between the laminations.

Make provision for protecting batts before, during and after construction.

Supporting and fixing the batts
The batts completely fill the cavity and are squeezed between two rows of ordinary wall ties spaced 450 mm apart vertically and 900 mm apart horizontally (staggered).

Beginning at ground level
The first row of batts may be required below ground level resting on the concrete cavity fill. If necessary cut the batts to fit tightly below the first row of wall ties. Batts may be cut with a trowel or a long knife *(fig 4.26a)*.

Alternatively, begin by supporting each batt on two ties in the row at DPC level *(fig 4.26b)*. Additional ties will be required at this level.

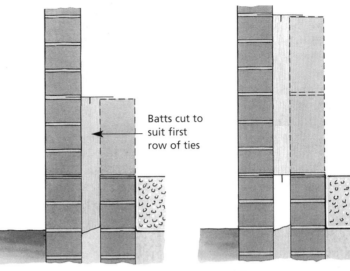

Batts cut to suit first row of ties

Figure 4.26a.

Figure 4.26b.

Remove excess mortar
Water can penetrate where fresh mortar from the bed joints in the outer leaf has squeezed into the cavity level with the horizontal joints between batts *(fig 4.27)*. The water will drain vertically down the laminations but further obstructions may deflect it towards the inner leaf.

Figure 4.27.

Raising the outer leaf first is preferred
The risk of rain penetration is minimised by building the outer leaf first.
- This enables the cavity side of the joints to be struck flush *(fig 4.28)*.
- Place a protective board to collect the mortar droppings and save time and effort later.
- Any mortar left on the top of the batts must be removed otherwise water may penetrate the joint.

Figure 4.28.

- If leading with the outer leaf, build to a height just sufficient to hold the next row of ties securely in place.

Raising the inner leaf first is second best
If the inner leaf has to be built first:

- After placing the batts build one course of bricks *(fig 4.29)*.
- Strike the mortar joint flush on the cavity side and clean off any mortar droppings.

- Place the batts in the trough which has been formed. Never push the batts into deeper cavities as mortar may be dislodged and bridge the cavity.

Cutting batts
It will sometimes be necessary to use small strips cut from batts. Do not place the end laminations of cut pieces (i.e. the cut surfaces) in contact with the external leaf. Otherwise they may conduct moisture to the inner leaf *(fig 4.30)*.

Fitting over extra ties
Where extra ties are required at window and door jambs cut a neat slit for them. Never tear the insulation or force it over the ties *(fig 4.31)*.

Gable walls
The insulation in gable walls should preferably go up to the verge. If it does not, the top of the insulation must be protected by a DPC cavity tray that has

stop ends to prevent water from the tray draining into the insulation *(see fig 4.24d)*.

2. BUILDING-IN PARTIAL-FILL CAVITY BOARDS

Materials
Insulation boards are rigid and fixed flat to the cavity face of the inner leaf. They may be made from expanded polystyrene bead board; extruded expanded polystyrene or polyisocyanurate foam, and glass fibre.

Make provision to store boards in the dry and on a *flat* surface.

Distorted boards will be difficult to build-in correctly.

Clear air space
The clear air space remaining after the boards have been fixed is usually not less than 50 mm. Good cavity wall techniques will keep the space clear of mortar droppings. Occasionally, in very sheltered areas, a 25 mm air space may be specified and this demands very great care from the bricklayer if the clear air space is not to be bridged *(fig 4.32)*.

Supporting and fixing the boards
The boards are supported between two rows of ties and held back tightly against the inner leaf by special clips. There must be at least two ties top and bottom and so the ties must be in vertical rows (not staggered) 600 mm apart horizontally. This is less than the maximum spacing allowed by the Building Regulations for structural purposes and is therefore acceptable.

Figure 4.29.

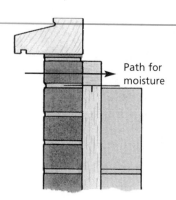

Path for moisture

Figure 4.30.

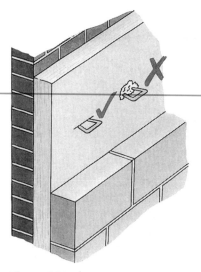

Figure 4.31.

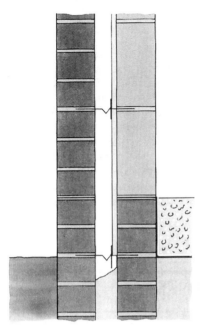

Figure 4.32. *Clear air space.*

Special ties

Specially designed retaining wall ties must be used and one example which has a plastic ring clip is illustrated *(see fig 4.24b)*.

Do not fix boards by any other method such as wedges or using the drips of butterfly ties.

Beginning at ground level

Whether the first row of boards is required below or above DPC, each board must be supported and retained by ties *(see fig 4.32)*.

Building sequence

- Build up the inner leaf and strike the joints flush on the cavity side.
- Remove droppings from the wall tops and prevent their getting into the cavity.

- Fit and fix the boards securely, staggering the vertical joints, and butt the horizontal and vertical joint closely *(fig 4.33a)*. Do not overlap the boards except in the case of those having specially rebated edges.
- Build up the outer leaf keeping the cavity clear of mortar droppings. Protect insulation and cavity before continuing the inner leaf *(fig 4.33b)*.
- Do not change fixing position from one leaf to another.

Extra ties

Cut neatly around ties – do not impale or break boards.

3. INJECTION OF INSULATION AFTER WALL CONSTRUCTION

The insulation is usually injected some six months after completion of the building to allow the structure time to dry. The bricklayers will not be directly involved as with boards and batts but careful workmanship is still essential if rain penetration is to be avoided. In particular it is essential to:

- Maintain the cavity width as specified.
- Keep wall ties clear of mortar droppings.
- Sleeve all vents and seal all openings.
- Provide a cavity tray at the ceiling level in any gable wall if the insulation is not to be carried up to the apex.

Above all seek the advice of the insulation contractor before building the wall.

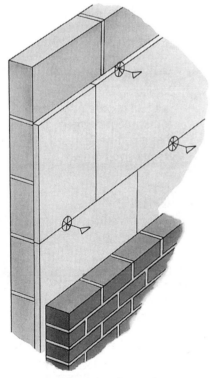

Figure 4.33a. *Fit and fix board securely.*

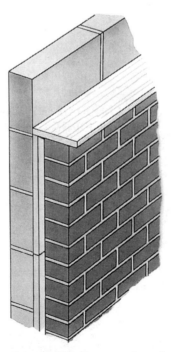

Figure 4.33b. *Protect cavity and insulation.*

4.5 VERTICAL MOVEMENT JOINTS

Effective movement joints are essential to prevent brickwork in modern buildings cracking from the inevitable movements which will occur over many years. Architects and engineers design movement joints from experience and the results of research, but they have to rely on bricklayers to build the movement joints with care and attention.

An understanding of the causes of movement and the prevention of cracking will help bricklayers to apply their skills more effectively.

THE CAUSES OF MOVEMENT

All materials expand and contract due to hourly, daily and seasonal changes in temperature (**thermal movement**) and wetting and drying (**moisture movement**).

In addition to these fluctuating and reversible movements, newly fired clay bricks expand continually over many years as they gradually take up moisture. On the other hand calcium silicate bricks (sandlime and flintlime), like concrete products, contract as they dry after being removed from high pressure steam autoclaves.

THE LOCATION OF MOVEMENT JOINTS

Designers try to avoid cracking in long runs of brickwork by dividing them into shorter lengths with vertical movement joints. These are generally at 10 to 15 m intervals in clay brickwork but may be as close as 6 m in calcium silicate brickwork. The joints are usually between 10 and 20 mm wide.

The location and form of movement joints may depend on the architect's decision to either minimise or emphasise their appearance (fig 4.34).

FORMING MOVEMENT JOINTS

If the movement joints are to function effectively and enhance rather than spoil the appearance of the building they must be built with care, attention and understanding. It is most important to maintain:

- Joints free from mortar and debris.
- Full bed joints to the face within the movement joint so that it can be sealed effectively and neatly.
- Correct and constant joint width.
- Verticality.
- Bed joints either side of the movement joint at the same level.

Two methods

There are two basic methods of forming movement joints:
1. The specified joint filler is positioned by suspending or bracing it plumb and in line with the intended face of the wall. It is then built-in (fig 4.35a).

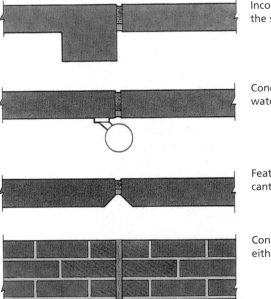

Inconspicuously located at the side of a pier

Concealed behind a rain water pipe

Feature made with single cant bricks

Contrasting coloured bricks either side of a joint

Figure 4.34. *Minimising and emphasising movement joints.*

The filler should initially be flush with the external face. Remove sufficient filler to provide the full specified depth and clean brick surfaces for the sealant to be applied and adhere effectively *(fig 4.35b)*.

An alternative method is to cut from the filler a piece equivalent to the specified sealant depth and then tack it back before building-in. The cut piece can then be easily removed in order to apply the sealant *(fig 4.35c)*.

It is essential that sufficient space be left for the specified depth of sealant. This will have been carefully calculated so that the sealant can adjust to the movements in the joint and remain effective for many years.

WARNING! There is now some evidence which suggests that when this method is used the mortar may squeeze from between the bricks during laying so that the hardened mortar compresses the filler and reduces the width and effectiveness of the movement joint. If there is any risk of this happening the second method described below would be preferable.

The joint filler is often a semi-rigid or closed cell polyethylene strip. Hemp, fibre board, cork and other similar materials must **not** be used in clay brickwork as they are insufficiently compressible to allow the joints to close as the brickwork expands.

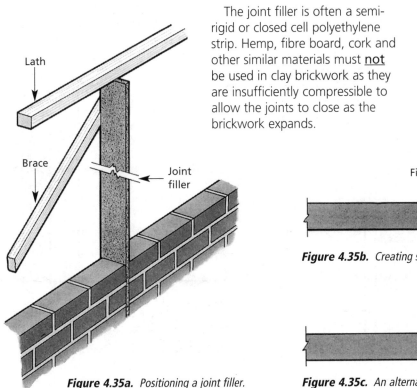

Figure 4.35a. *Positioning a joint filler.*

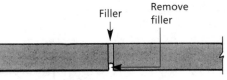

Filler   Remove filler

Figure 4.35b. *Creating space for sealant.*

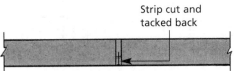

Strip cut and tacked back

Figure 4.35c. *An alternative method.*

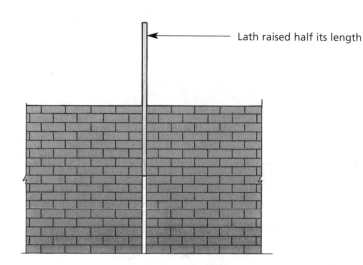

Figure 4.36a. *Forming an open joint.*

Figure 4.36b. *Raising the lath.*

2. The joints are formed by building-in a temporary timber lath for the full depth of the brick skin and the width of the joint. The lath should be braced in position slightly behind the brickwork face to prevent its fouling the bricklayer's line *(fig 4.36a)*.

The lath should be 'tapped' from time to time to break the mortar bond so that it may eventually be removed without damage to the brickwork. The lath should be checked frequently for plumb.

When the brickwork reaches the top of the lath, raise the lath about half its length in order to maintain it plumb with the minimum of bracing *(fig 4.36b)*.

All mortar droppings and debris must be removed from the joint before it is filled and sealed.

If the length of a wall requires that it be divided into 'manageable' lengths for building, the use of a temporary profile provides a true and plumb surface for the joint filler and sealant whilst enabling the wall to be run-in for line level and gauge *(fig 4.37)*.

The subsequent section of wall may be built by 'pinning into' the

Figure 4.37. *A temporary profile.*

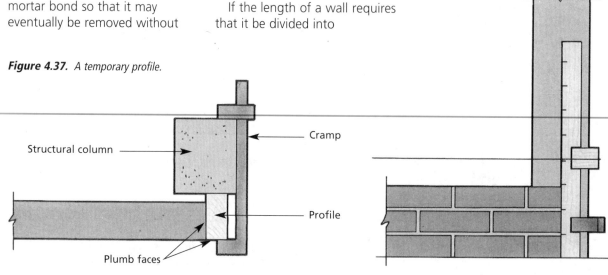

Plan

Elevation

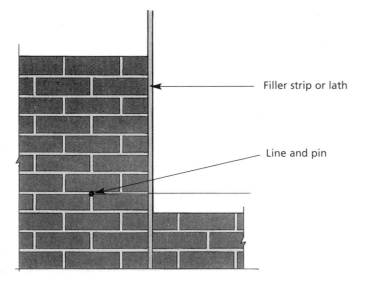

Figure 4.38. *The next section.*

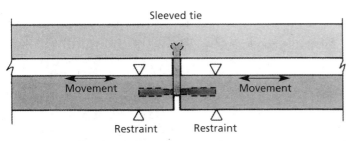

Figure 4.39a. *Special ties restrain walls but allow movement.*

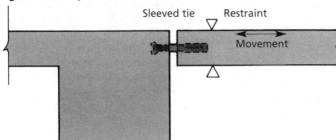

Figure 4.39b. *Ties as delivered for use with concrete or calcium silicate bricks and blocks.*

existing wall for line and gauge whilst the filler strip or lath can be placed securely against the end of the wall *(fig 4.38).*

BUILDING-IN SPECIAL TIES ACROSS MOVEMENT JOINTS

Vertical movement joints allow brick walls to move horizontally along their length but it is necessary to prevent the ends of the leaves on both sides of a movement joint from moving at right angles to their length. This is often done by tying them to a steel or reinforced concrete frame or to brick or block walls with special ties which have plastic debonding sleeves *(fig 4.39a).*

The ties are delivered ready for use with concrete blocks or calcium silicate bricks *(fig 4.39b).*

For use in clay brickwork the ties must be prepared by pulling them from the sleeves by an amount equal to the specified joint width to allow for expansion of the brick panels *(fig 4.39c).*

Figure 4.39c. *Ties as prepared for use with clay bricks.*

APPLICATION OF SEALANTS

After forming the joints, equal care must be taken in applying the sealant, preferably by skilled and experienced specialists. However, as the bricklayer is often required to apply the sealant some brief guidance is given here.

When applying sealants, the golden rule is to follow the manufacturer's instructions exactly.

The sealant must adhere to the sides or opposing faces of the joint but **not** to the filler or backing material *(fig 4.40)*. Most backing materials do not promote adhesion. Where they do a breaker strip should be used.

Types of sealant

Sealants may be ready-to-use in cartridge form or be supplied in two parts to be mixed on site. Both types are gun applied. Two suitable types are:

1. **Polysulfide sealants**
(i) One-part sealants in cartridges with a curing time of 2–3 weeks.
(ii) Two-part sealants requiring mixing prior to use and having a shorter period for

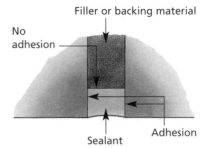

Filler or backing material

No adhesion

Sealant

Adhesion

Figure 4.40. Adhesion requirements.

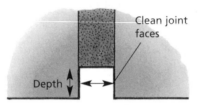

Clean joint faces

Depth

Figure 4.41. Preparation of joint.

application but curing more rapidly within 1–2 days.

2. **Silicone sealants**
A one part sealant which cures rapidly on exposure to air.

Joint preparation

The filler or backing material should be positioned to give the correct depth for the sealant *(fig 4.41)*.

The joint faces must be cleaned free of loose particles, release agents or water

repellants. The joint faces should be primed, if required by the sealant manufacturers.

Applying sealant

Two-part sealants must be mixed thoroughly.

The sealant should completely fill the joint and it should be 'tooled' as required by the manufacturer to compact the sealant, improve its adhesion to the joint faces and give a consistent, smooth, slightly concave finish. It is good practice to fix masking tape either side of the joint to minimise contamination of the brick face.

KEY POINTS

- Keep joints clear of mortar and debris.
- Maintain the specified width.
- Fill the bed joints flush within the movement joint.
- Provide the specified depth to receive the sealant.
- When applying sealants follow the manufacturer's instructions exactly.
- Pull the sleeves of movement ties beyond the end of the tie when using them with clay bricks.

4.6   REINFORCED AND POST-TENSIONED BRICKWORK

The designers of buildings and civil engineering structures are constantly using new techniques and materials for greater strength and economy.

Normally, unreinforced brickwork is strong enough to carry loads which bear directly downwards on it, e.g. a floor resting on a wall *(fig 4.42)*. Such loads try to crush the wall which is said to be in compression.

Lateral loads, like the wind, try to bend the brickwork so that part is in tension. Brickwork is strong in compression but weak in tension, so something must be done to resist the tension or else it will crack and fail *(fig 4.43)*.

This section shows how steel reinforcement can

LOAD

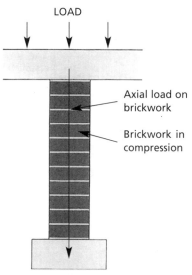

Axial load on brickwork

Brickwork in compression

Figure 4.42. *Brickwork in compression from axial load.*

LOAD TO ONE SIDE OF WALL, i.e. eccentric

Wind load

Brickwork in tension this side

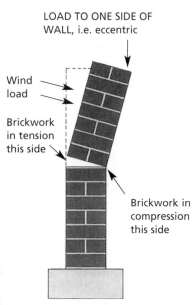

Brickwork in compression this side

Figure 4.43. *Brickwork in tension from lateral and eccentric loads.*

strengthen brickwork to resist tension.

As steel reinforced brickwork is becoming more common it is important that bricklayers know why and understand the need for good construction practice.

Figure 4.44. *Bed joint reinforcement.*

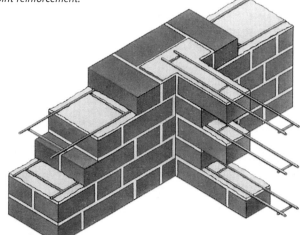

1. BED JOINT REINFORCED BRICKWORK

Steel reinforcement is laid in the bed joint mortar to control movement in the brickwork, to tie leaves of brickwork together and increase brickwork strength *(fig 4.44).*

2. REINFORCED BRICKWORK BEAMS

Reinforced brick beams can be created by building 'U'-shaped brickwork and then placing steel and infill concrete in the void.

3. GROUTED CAVITY REINFORCED BRICKWORK

Either steel rod or steel mesh reinforcement is placed in the centre of the cavity. The cavity is then filled with either mortar or, more commonly, a reasonably liquid concrete or grout *(fig 4.45).*

Figure 4.45. *Grouted cavity reinforced brickwork.*

Reinforcement

Cavity fill

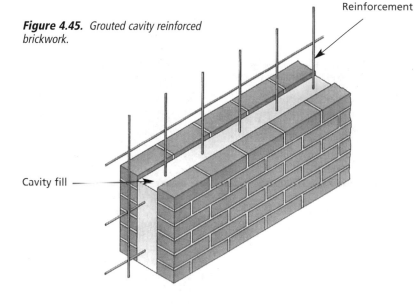

The purpose of filling the cavity is to bind the reinforcement to the brickwork creating a solid structure. Cavity fill also protects the steel from corrosion. The amount of protection required depends on the type of steel and the exposure of the particular structure.

The technique enables brickwork to resist lateral loads, such as wind, in buildings and freestanding walls. It is also used in earth-retaining walls normally not exceeding 2 m high.

4. QUETTA BOND REINFORCED BRICKWORK

Steel is contained within pockets formed by the brickwork bonding *(fig 4.46)*. The pockets are normally filled with mortar which bonds the brickwork and steel together. Grouted cavity construction is more popular than Quetta bond as the latter is more difficult to build.

5. POCKET REINFORCED BRICKWORK

Reinforcement is contained within pockets in the rear of the wall between 'T'-shaped brickwork *(figs 4.47 and 4.58)*. The reinforcement is joined to the brickwork by the concrete infill forming a solid structure. Pocket walls can be faced on one side only and, therefore, are normally used for retaining walls which have only one side visible.

6. POST-TENSIONED BRICKWORK

Special techniques and steels are used to apply a permanent compressive load to the

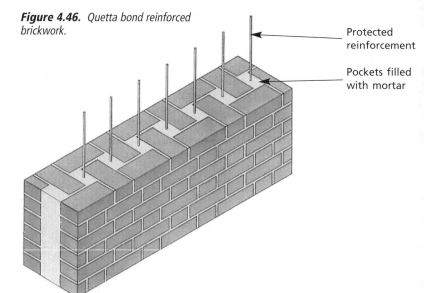

Figure 4.46. *Quetta bond reinforced brickwork.*

Protected reinforcement

Pockets filled with mortar

Figure 4.47. *A pocket-type retaining wall under construction.*

brickwork making it more capable of resisting horizontal loads. This technique can be applied to many forms of brickwork such as cavity, diaphragm and fin walling *(fig 4.48)*. It is also used in civil engineering structures such as bridge abutments and large retaining walls.

THE APPLICATION OF REINFORCED AND POST-TENSIONED BRICKWORK

Reinforcing and post-tensioning allows brickwork to carry greater loads in the same way as does

reinforcing and pre-stressing concrete. Both reinforcing and post-tensioning are particularly useful when brickwork has to resist lateral loads, e.g from wind, retained earth, and impact loads *(fig 4.49)*. Because both reinforced and post-tensioned brickwork can carry greater loads than unreinforced brickwork, which would have to be considerably thicker, they are often cheaper to build.

CONSTRUCTION TECHNIQUES

As both reinforced and post-tensioned brickwork are designed by engineers to BS 5628:Part 2:1985[1], often to carry heavy loads, they must be very carefully built.

1. HORIZONTAL REINFORCEMENT

Honzontal reinforcement is normally used to:

a) help control movement in long runs of brickwork.

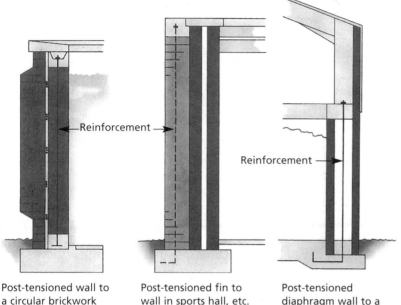

Post-tensioned wall to a circular brickwork water tank

Post-tensioned fin to wall in sports hall, etc.

Post-tensioned diaphragm wall to a grainstore

Figure 4.48. *Some applications of post-tensioned brickwork.*

Figure 4.49. *Bridge parapet wall reinforced to withstand vehicle impact – so hope the ducks!*

Figure 4.50. *Unrolling bed joint reinforcement.*

b) enable brickwork to resist horizontal loads, e.g. the wind.
c) provide a reinforced brick beam either to spread a point load from say the end of a steel beam over a larger area of brickwork or to create a reinforced brick beam over an opening.

Bed joint reinforcement
The simplest form of horizontal reinforcement is bed joint reinforcement *(fig 4.50)*. The reinforcement may be placed in every third or sixth mortar bed joint depending on engineering considerations.

When using bed joint reinforcement note the following:

a) Reinforcement wires should be no greater than 6 mm in diameter otherwise it will be difficult to fit the steel into a 10 mm bed joint.

b) Place reinforcement in a bed of mortar with a minimum of 15 mm between the face of the brickwork and the edge of the reinforcement. The steel must be surrounded by mortar to develop strength and protect galvanised steel from corrosion.

c) Some steel manufacturers provide special jointing sections to avoid having to lap the steel bars where the reinforcement strips meet. Occasionally it may be necessary to cut some of the cross wires to achieve a successful 150 mm lap between adjoining sections of steel. This is not good practice and only the minimum amount of steel should be removed, and where possible the cross wires should be maintained as near as possible to the original spacing.

d) Normally, suitable stainless steel reinforcement is used for external solid walls and both leaves of external cavity walls. Galvanised steel or bitumen-coated steel is normally used only for internal walls.

e) When building-in special bed joint reinforcement to tie two leaves of brickwork together across a cavity, build up both leaves together, otherwise it will be difficult to align the steel between the two leaves.

f) Do not build-in reinforcement across movement joints. The steel will stop the movement joints working.

g) The majority of bed joint reinforcement is provided as 'flat' steel, but if coiled reinforcement is being used *(fig 4.50)* roll it out carefully to avoid stressing the wires or damaging them in some other way.

h) Make small marks on the faces of mortar joints to show which contain reinforcement.

i) Extend bed joint reinforcement the specified length within the brickwork, particularly if it is reinforcing brickwork over openings. Bed joint reinforcement is normally used only for light reinforcement.

Other horizontal reinforcement

Stronger forms of reinforced brickwork are used for lintels or beams over openings. Normally, reinforcement will be contained in a void within the brickwork and will be bonded or joined to the brickwork by infill concrete or grout *(fig 4.51)*.

Whichever form of reinforcement is used:

a) Provide a temporary support for the brickwork until it has cured sufficiently to carry the loads over the opening. This is normally between 14 and 28 days after the infill concrete has been placed. Make sure the temporary support is strong enough to carry the loads which will rest on it during construction.

b) Prevent mortar squeezing from the joints during bricklaying by placing foam or rubber strips on the shutter at the base of joints *(fig 4.52)*.

c) Because the reinforced brickwork will be built in a stronger mortar mix, ensure a constant colour of mortar throughout the brickwork by raking back all the joints in the reinforced brick section and then pointing up with matching mortar once the temporary support has been removed.

d) Do not build further brickwork over the opening until the reinforced brickwork section has cured sufficiently to carry the imposed load.

e) Check the tolerances of the bricks before building them into the brickwork. As they rest on the temporary support, any variation in the brick length will show at the top edge of the bricks *(fig 4.51)*.

2. VERTICAL REINFORCEMENT

Brickwork can be vertically reinforced to increase its bending strength.

For example, a gate hung on a tall brick pier will try to pull it over *(fig 4.53)*. To resist bending and cracking in the brick pier, vertical reinforcement can be built into the centre.

The mass of earth behind an unreinforced retaining wall may cause the brickwork to fail in tension. Steel placed in the wall can help resist the load by preventing the wall sliding forward or bending and cracking.

Vertical reinforcement can be incorporated in brickwork in several ways. Each has its own construction method:

Grouted cavity reinforced brickwork *(see figs 4.45 & 4.54)*

a) Construct reinforced concrete foundation with reinforcing starter bars typically 600 mm high, projecting from the top face. Lap reinforcement in the grouted cavity brickwork over the starter bars to join the wall to the foundation *(fig 4.55)*.

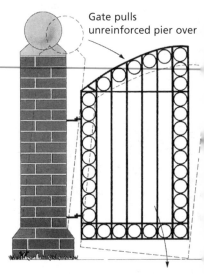

Gate pulls unreinforced pier over

Figure 4.53. *A gate pier.*

Figure 4.51. *Forming pocket for horizontal reinforcement.*

Figure 4.52. *Foam strips placed on shutter to prevent mortar squeezing out.*

Figure 4.54. *Grouted cavity wall under construction – note loop ties.*

b) Build one leaf of brickwork off the foundation incorporating wall ties as specified. Do not build the leaf higher than 16 courses otherwise it may fall over.

c) Clean all mortar snots off the rear face of the brickwork and off the ties and the foundation area. Take care to avoid injury on the projecting wall ties.

d) Lap the reinforcement against the starter bars by the length specified (fig 4.55). Support the reinforcement within the cavity, typically by tying the reinforcement to the wall ties.

If ordinary mild steel reinforcement is used with stainless steel wall ties do not allow the two steels to touch, otherwise bi-metallic corrosion can occur. A non-metallic material should be used to fix steel reinforcement to wall ties.

Place steel centrally in the cavity except where it is lapped to adjoining steel. Plastic spacers help to do this.

Steel is usually placed in the centre of cavities to ensure that there is adequate infill concrete cover to the steel to prevent corrosion of unprotected reinforcing steels and to provide bond between the infill concrete and the steel.

e) Construct the other leaf of brickwork 6 courses high, removing all mortar droppings from the cavity. Use cavity battens to prevent mortar entering the cavity or place a piece of polythene in the base of the cavity to catch the mortar droppings and then when the brickwork is completed pull the polythene out of the cavity complete with the droppings (fig 4.55).

f) When the mortar has started to set and the brickwork is strong enough to resist the pressure of the infill concrete, typically after 12 to 15 hours, fill the cavity with concrete as specified. Make sure no voids are left in the infill concrete using a small poker vibrator to force the air out (fig 4.56).

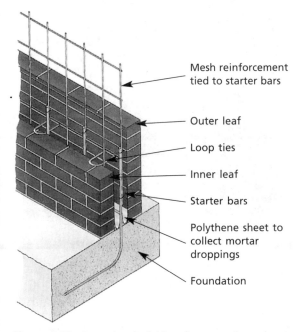

Mesh reinforcement tied to starter bars

Outer leaf

Loop ties

Inner leaf

Starter bars

Polythene sheet to collect mortar droppings

Foundation

Figure 4.55. *Grouted cavity brickwork construction – showing a method of collecting mortar droppings in cavity.*

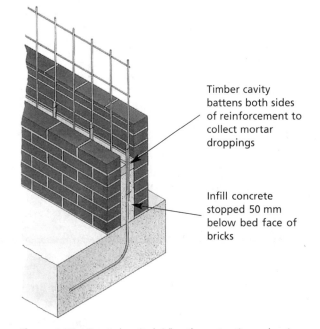

Timber cavity battens both sides of reinforcement to collect mortar droppings

Infill concrete stopped 50 mm below bed face of bricks

Figure 4.56. *Grouted cavity brickwork construction – showing an alternative method of collecting mortar droppings in cavity.*

Finish the concrete face 50 mm below the bed face of the lower leaf of brickwork to avoid a straight line of weakness through the brickwork *(fig 4.57)*. Take care that the infill concrete is not over-vibrated and separates out.

g) Repeat this operation ensuring mortar droppings are cleared from the grouted cavity section before the infill concrete is placed. Where there are several sections of grouted cavity wall being built it may be necessary to provide temporary ends to the cavity to prevent the infill concrete escaping.

It is important that any face brickwork is protected while infill concrete is placed, to avoid staining the face of the brickwork.

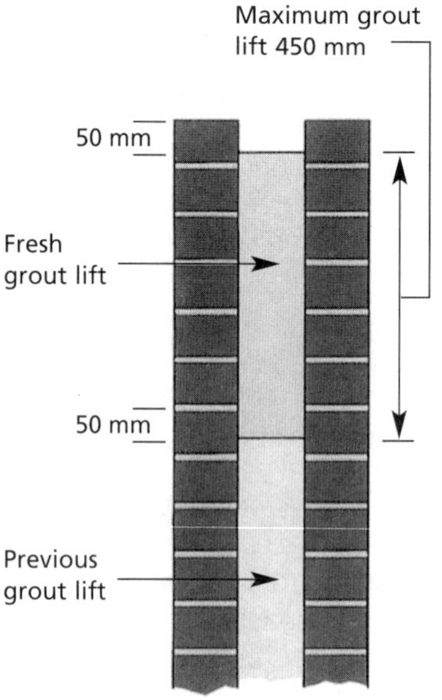

Quetta bond reinforced brickwork
(see figure 4.46)
a) Lay the bricks carefully to ensure a constant size of pocket around the vertical reinforcement. Incorporate starter bars at the base of the wall to join the vertical reinforcement within the wall to the foundation.
b) The spaces between the steel and the brickwork are often filled with Designation (i) mortar as bricklaying progresses. It must be well compacted around the steel which will have to be protected either as galvanised steel or as stainless steel. With mild steel, concrete is normally used to fill between the steel and the brickwork.

Because the void in which the infill concrete has to be

Maximum grout lift 450 mm

50 mm

Fresh grout lift

50 mm

Previous grout lift

Figure 4.57. *Grouted cavity construction – placing grout in low lifts.*

placed is narrow, take care to ensure that the concrete does not hang on the side of the void but is well compacted in place.

Pocket reinforced brickwork
(figure 4.58)
a) For ease of construction the brickwork is normally completed before the reinforcing steel is placed.
b) Keep all mortar out of the pocket and take particular care with the bonding around the pocket to ensure the sides of the pocket are built into the main brickwork as specified.
c) When the brickwork is complete, place the

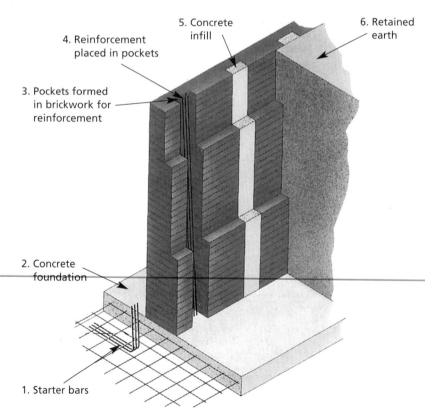

5. Concrete infill

6. Retained earth

4. Reinforcement placed in pockets

3. Pockets formed in brickwork for reinforcement

2. Concrete foundation

1. Starter bars

Figure 4.58. *Pocket reinforced brickwork – showing normal sequence of building operations. (The damp-proof membrane to the retaining face has been omitted for clarity.)*

reinforcement steel in the pocket lapping over the starter bars at the base. Fix shuttering at the rear of the pocket prior to the infill concrete being poured into the void.

Vibrate the concrete into place, taking care not to disturb the brickwork. This ensures that the infill concrete forms a solid mass around the steel within the brickwork pocket.

POST-TENSIONED BRICKWORK
(figure 4.48)

a) Larger diameter reinforcing bars are normally used for post-tensioned brickwork. These may require a temporary support to stop them bending and damaging the green brickwork during construction.
b) The bars are often contained in voids, cavities or ducts within the brickwork. Even though the post-tensioned steel is often left ungrouted within the void, cavity or duct it is still important that all mortar droppings are cleared from the void around the steel.
c) If specified, construct sloping concrete plinths at the base of the void around the post-tensioning bars and form weep holes through the brickwork to allow the water to drain.
d) At the top of the post-tensioned brickwork wall a reinforced concrete capping beam will normally be constructed to spread the load produced by the tensioning of the

reinforcement *(fig 4.59)*. This enables the force produced by the tensioning to be spread evenly over the brickwork.

The reinforced concrete beam may be cast *in situ*. Otherwise it will be preformed, in which case bed it on the brickwork on full mortar joints.
e) Tension the reinforcement after the specified period, not less than 14 days. Retension 14 days later to allow for any settlement which may have occurred in the brickwork due to the load imposed on it.

GENERAL POINTS
Whichever form of reinforced or post-tensioned brickwork is used, certain bricklaying operations are important:

a) Locate the steel carefully in the brickwork in accordance with the engineer's drawings and specification.
b) Take care to support the steel during bricklaying so that steel does not cause injury to the bricklayers or damage the green brickwork. In particular, cover the ends of rods to prevent eye injuries.
c) Fill all joints in the brickwork completely. The brickwork will normally carry very high loads.
d) Mix the mortar accurately and use quickly as the higher cement content will lead to a faster set. Reinforced brickwork normally requires higher strength mortars such as Designation (i) (1:¼:3) and Designation (ii) (1:½:4½).

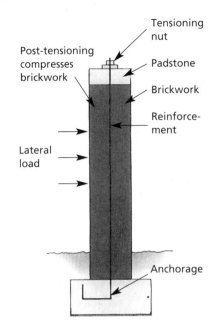

Figure 4.59. *The principle of post-tensioned brickwork.*

e) Accurately form reinforcement lap joints as specified.
f) Remove all excess mortar from the reinforcement and ties from the voids. Mortars are weaker than the infill concrete and should not be mixed with it.
g) When placing infill concrete, cover the face brickwork to protect it from concrete splashing. Be careful not to overfill with grout as spillage is difficult to remove from the face brickwork.
h) Damp-proof membranes will often be brushed, sprayed or stuck on the rear face of reinforced brickwork retaining walls. In which case fill all joints on the rear face and finish with a firm surface. Unless otherwise instructed form a shallow bucket handle profile for the rear face.
i) Reinforced and post-tensioned brickwork must be cured

properly and protected from saturation during construction. *(see Section 1.2 'Protection of newly built brickwork')*

properly designed and installed copings or roof overhangs, and damp-proof membranes on the rear surface of retaining walls.

Reference

(1) BS 5628:Part 2:1985 'Structural use of reinforced and prestressed masonry'.

A SPECIAL NEED FOR TEMPORARY PROTECTION

When construction of a section of reinforced brickwork has been completed, ensure that temporary protection is immediately provided to prevent rainwater percolating through concrete or mortar infill and then to the facing brickwork, as this can lead to serious lime staining. Design of the completed structure should have features that give permanent protection against such wetting, e.g.

KEY POINTS

- ■ Position steel correctly in the brickwork.
- ■ Keep mortar droppings clear of brickwork ties and the base of voids around the reinforcement.
- ■ Properly form steel laps.
- ■ Support steel to avoid injury to bricklayers and damage to green brickwork.
- ■ Use only specified mortar and mix properly.
- ■ Completely fill all mortar joints and finish as specified.

- ■ Protect facework during placing of infill concrete.
- ■ Protect all brickwork after construction and during inclement weather.
- ■ Reinforced and post-tensioned brickwork is an engineering structure and the highest standards of brickwork are required.
- ■ Do not subject brickwork to loading other than self-weight until it has cured sufficiently.

4.7   BRICKWORK ON METAL SUPPORT SYSTEMS

This section deals with buildings above two storeys in height that have the brickwork outer leaves of cavity walls supported by a system of metal angles and/or brackets fixed to structural frames.

Bricklayers may themselves fix a system or build on one fixed by others. Either way, systems should be fixed strictly to the designer's specification and manufacturer's instructions.

If you are ever in doubt, ask a supervisor. Mistakes are usually expensive to put right and may be dangerous.

VERTICAL SPACING OF SUPPORT SYSTEMS

Systems are normally located at every storey or second storey *(fig 4.60)*. They are designed specifically to support the brickwork for each particular building.

Concrete frames shrink as they cure and creep under load. Steel frames are considered dimensionally stable. Because clay brickwork slowly expands for a long time, horizontal movement joints are normally positioned between the top of the brickwork and the underside of each metal support system,

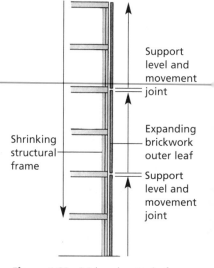

Figure 4.60. Brickwork outer leaf supported at every second floor.

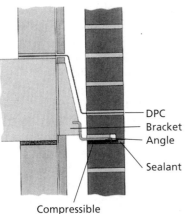

Figure 4.61. Typical detail of a support system.

whether the structural frame is of concrete or steel. The joints, which are usually designed to accommodate an expansion of brickwork of up to 1 mm per metre, will therefore be compressed over a period of time *(fig 4.61)*.

TYPES OF METAL USED FOR SUPPORT SYSTEMS

For buildings of more than three storeys, the support system will always be manufactured from austenitic stainless steel. Although galvanised mild steel is permitted in buildings of three storeys and less, in practice, the building specification or local regulations will invariably call for austenitic stainless steel which is corrosion resistant with a very long maintenance-free life. Galvanising is a sacrificial coating, has a limited life, and the unprotected mild steel will corrode and may damage the brickwork.

The specification for austenitic stainless steel is normally a minimum 18/8 composition (18% chromium, 8% nickel); Grade 304 is the most popular of the 18/8 stainless steels. That for galvanised steel, if used, is normally Grade 43 mild steel with a minimum post galvanised coating of 940 g/m^2.

FIXING DISSIMILAR METALS TOGETHER

When dissimilar metals are in contact and moisture is present, electrolytic action increases the corrosion rate of the less noble material, e.g. when stainless and mild steel are in contact, the corrosion rate of the mild steel increases.

In many cases the protective coating provided to the mild steel may be sufficient to prevent this action. Where this is not the case the designer may decide to prevent electrolytic action by one of two methods.

1. Isolation of two metals.
Impervious packing, sleeves and washers separate stainless steel from mild steel *(fig 4.62)*.

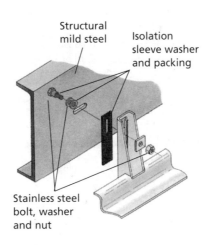

Structural mild steel

Isolation sleeve washer and packing

Stainless steel bolt, washer and nut

Figure 4.62. A typical method of isolating two dissimilar metals.

2. Exclusion of moisture.
Jointing compound and a paint system overlapping the joint by a minimum of 20 mm may also be specified. Thick water-resisting plastics or mastic coatings should be used.

TYPES OF SUPPORT SYSTEMS
Brickwork support systems are usually selected from one of three types:

1. Continuous angle systems.
Lengths of continuous angle of up to 3 m long, with horizontal slotted holes along the vertical leg, are fixed to concrete with either site-drilled expansion bolts or serrated 'T'-head bolts into vertical cast-in toothed channels *(fig 4.63)*.

10 mm open joints are left between adjoining lengths of angles to allow for thermal expansion of the angle as well as building tolerances.

Angles may be bolted directly to steelwork with stainless steel nuts and bolts. (See heading 'FIXING DISSIMILAR METALS TOGETHER'.) Examples are given in the BDA/BSC joint publication 'Brick cladding to steel framed buildings'.[1]

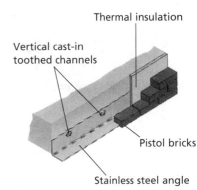

Thermal insulation

Vertical cast-in toothed channels

Pistol bricks

Stainless steel angle

Figure 4.63. Continuous angle systems. (Cavity tray omitted for clarity.)

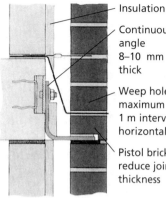

Figure 4.64. *Continuous angle systems.*

Depending on the mass of brickwork to be supported and the cavity width specified, the angle may be typically 8 to 10 mm thick. Pistol bricks may be used to avoid an excessively thick joint in face of the wall *(fig 4.64)*.

2. Bracket angle systems.
Much smaller angles are welded to brackets to suit a particular cavity width. The brackets are normally fixed with 'T'-head bolts into a continuous horizontal channel cast into concrete *(fig 4.65)*.

The angle thickness will normally be 5 mm or 6 mm. It may sometimes be built into a joint

without using a pistol brick but the joint will be wider than 10 mm.

3. Individual bracket systems.
Bricks are supported on individual brackets which have stiffeners which fit into the vertical cross joints. The brackets are fixed in a similar manner to bracket angles *(fig 4.66)*.

The support bases are usually 4 mm thick and are built into a nominal 10 mm joint.

Summary of applications
The three types of systems described previously are typically used as follows.

Continuous angle
- Usually for cavities less than 75 mm.
- To close the cavity where the underside of the support will be seen.

Bracket angle
- For cavities from 50 mm to 150 mm.
- For fixing directly to uncased structural steelwork.

Individual brackets
- For brickwork that is curved on plan.
- For features such as arches and suspended soldier courses.

BUILDING WITH CONTINUOUS ANGLE OR BRACKET ANGLE SYSTEMS

Accurate positioning
Angles should be correctly positioned, levelled and securely fixed to the frame. There should be room underneath angles to allow for vertical expansion of the brickwork below and shrinkage of any concrete frame.

Where 10 mm joints are left between lengths of adjoining continuous angle, in order to allow for tolerances and thermal expansion, it may be necessary to seal them with a local DPC to prevent water crossing the cavity. Joints in bracket angle systems may be left open.

Fixing
All fixings should be tightened to the specified torque and the overall thickness of any shims should not exceed the supplier's recommendations, usually 12 mm or 20 mm *(fig 4.67)*. Shims should be of the 'horseshoe' type, giving support to the full depth of the back of the angle including the heel

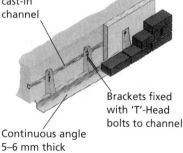

Figure 4.65. *Bracket angle system. (Cavity tray omitted for clarity.)*

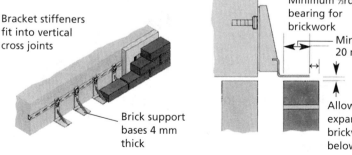

Figure 4.66. *Individual bracket system. (Cavity tray omitted for clarity.)*

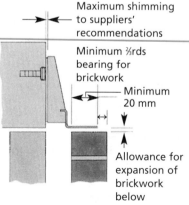

Figure 4.67. *Positioning and shimming angle.*

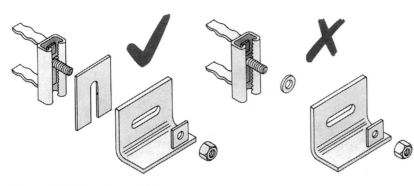

Figure 4.68. *Use of correct shims.*

(fig 4.68). Round washers must not be used as they will allow the angle or bracket to deflect.

Building the brickwork
Designers' views differ on the detailed solution to this relatively recent technique of supporting outer leaves of brickwork on steel angles. Consequently, bricklayers may be required to position the DPC either on the angle or one course above. A typical method of building each is described below.

1. DPC above the first course.
(fig 4.69)
Bed the first course on the angle. Use a reduced mortar bed if instructed to keep the overall joint thickness to a minimum. Rebated bricks (sometimes called pistol bricks) may be specified.

In order to improve the bond between the first course and the angle some specifiers may require a polymer bonding admixture to the mortar, such as styrene-butadiene rubber (SBR) in which case it must be used strictly in accordance with the manufacturer's instructions. Others may have specified a profiled surface to the top of the angle or mesh to be tack welded to it.

Continue building the outer leaf including the DPC and ties to the inner leaf. Because this method isolates one course between the angle and the DPC, some designers specify ties, fixed back to the support system but adjustable horizontally, for building into cross joints in the first course *(fig 4.69)*.

2. DPC on support angle.
Normally, DPCs will be bedded on the angle before bedding the first course although some designers may require the DPC to be laid directly on the angle to provide a slip plane between it and the brickwork which have different movement characteristics. It is likely to be simpler to hold a flexible DPC in place while bedding the bricks, rather than a thicker, stiffer material.

The bricks for the first course may be rebated in order to minimise the visible joint thickness.

Whatever the position of the DPC ensure that open cross joints or weep holes are formed at a maximum of 1 m intervals immediately above the DPC. *(see Section 4.3 'Damp-proof courses')*

Half-brick leaves should have a minimum bearing on the angles of ⅔ the width of the leaf, approximately 70 mm. The toes of angles should be at least 20 mm back from the face of the brickwork *(fig 4.67)*.

BUILDING WITH INDIVIDUAL BRACKET SYSTEMS

Positioning and fixing brackets
Unlike continuous angle and bracket angle support systems, individual brackets cannot be preset ready to lay bricks. The following procedure is typical.

- Set out brackets at brick centres and bolt **loosely** to the cast-in channel *(fig 4.70)*.
- Place bricks on brackets and adjust the latter so that the stiffeners are in the centre of

Figure 4.69.

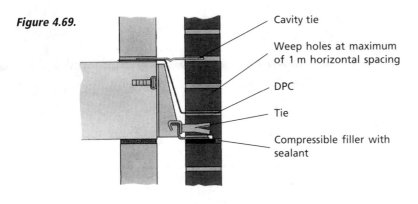

Cavity tie

Weep holes at maximum of 1 m horizontal spacing

DPC

Tie

Compressible filler with sealant

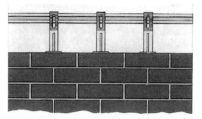

Figure 4.70. *Individual brackets in position.*

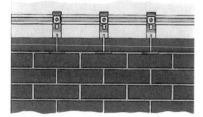

Figure 4.71. *Individual brackets levelled and tightened.*

cross joints and align each bracket and tighten the 'T'-head bolt to the correct torque *(fig 4.71).*

Building the brickwork
- A thin bed of mortar is normally applied to each bracket in order to align and level the bricks.
- Build the remaining brickwork including any DPCs and wall ties.

It may be advantageous to build the first course off the brackets using bricks selected for consistent length and height.

BUILDING A SOLDIER COURSE OVER AN OPENING

Bracket systems are often used to suspend brickwork over windows and openings when the architect does not want a lintel or support angle to be visible.

The individual brackets are positioned and tightened as described previously *(fig 4.72).*

A temporary support is fixed firmly to carry the weight of the soldier course, without movement, until the mortar sets.

The soldier bricks are perforated with at least two holes approximately 25 mm in diameter.

The soldier course is built from one end, three bricks at a time, completely filling both the holes and the joints with mortar. Short lengths of stainless steel stitching rods are pushed through the mortar holes to span between two brackets *(fig 4.73).*

The soldier course is continued in sets of three bricks until complete (fig 4.74).

A similar method can be used to build arches that cannot be designed to be self-supporting. The brackets are fixed to cast-in channels in the face of the concrete, radiating around the arch.

It is necessary to prevent mortar squeezing from the vertical joints and under the shutter/temporary support and staining the face of the bricks. Foam rubber strips may be placed in the bottom of each vertical joint and when the vertical joint is formed the face of the vertical joint should be raked back.

When the brickwork is cured, say 14 days after construction depending on the load imposed on the brickwork and the weather conditions, the temporary supports are removed, the foam strips removed and then the vertical face and the horizontal soffit face of the joint pointed up in matching mortar. Bricks selected for consistency of size will normally be required for this form of construction and the hanging system and the perforation pattern of the bricks must be compatible.

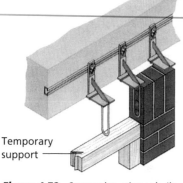

Temporary support

Figure 4.72. *Suspension stirrups built into vertical cross joints.*

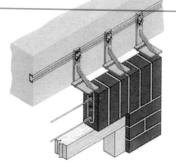

Figure 4.73. *Stitching rods pushed into mortar filled perforations.*

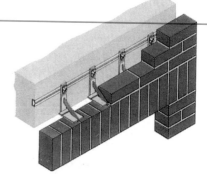

Figure 4.74. *Brick leaf built off individual brackets.*

BUILDING-IN WALL TIES

The need for correct, properly built-in cavity wall ties at the specified spacing was pointed out in section 4.2 'Ties in cavity walls'.

The need is even more critical when brick panels are supported on metal angle systems. Any deflection of the structural frame or support system may be transmitted to the brickwork leaf which should be restrained by being tied to the inner leaf and/or the structural frame.

In addition, the top of cavity wall panels may be tied to the frame just below the horizontal movement joint, for lateral support.

A typical tie, giving lateral restraint but allowing vertical differential movement, is shown in *fig 4.75*.

BUILDING 'CORBELLED' BRICKWORK*

True corbelling, as described in section 5.8, cannot be used in most modern cavity wall situations. Normally, if a corbelled appearance is required a support system is used, often with inverted plinth bricks *(fig 4.76)*.

The first course is built on a support system as previously described. The remaining corbel courses are tied back to the structure at every bed joint.

Depending on the extent of the corbelling, the number of courses built at one time should

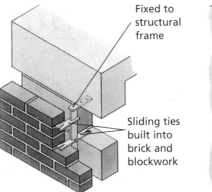

Figure 4.75. *Typical sliding restraint ties.*

Fixed to structural frame

Sliding ties built into brick and blockwork

be limited to possibly only two. The brickwork above the corbelled courses is carried on one of the support systems that have been described. Plinth bricks invariably require considerable modification to suit a system. Temporary support of the plinths during construction and measures to prevent mortar staining of the bricks and consistency of mortar colour must be taken.

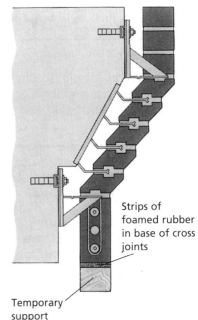

Strips of foamed rubber in base of cross joints

Temporary support

Figure 4.76.

Reference

(1) R. E. Bradshaw, G. Buckton, S. W. Southwick. 'Brick cladding to steel framed buildings'. The Brick Development Association and British Steel Corporation. September 1986.

KEY POINTS

- Check that the structural frame is within specified tolerances before fixing a support system.
- Ensure that any shimming does not exceed the maximum specified.
- Ensure that approved full-depth shims are used, not just round washers.
- Check that the support is at the correct level and that there is

- sufficient room underneath to allow for an expansion joint.
- Check that all bolts are tightened to the correct torque setting.
- Ensure a minimum of 70 mm–75 mm bearing for the brickwork.
- Build-in the correct cavity and restraint ties as specified.
- Follow manufacturers' instructions – if in doubt ask.

---

* Figures 7 and 11 inside the rear cover show good examples of corbelled brickwork.

5 SPECIFIC CONSTRUCTIONS

Some brickwork features require specific knowledge and skill to facilitate their construction. This section gives detailed guidance on a variety of construction features which, between them, illustrate the principles of several special procedures.

5.1 COPINGS AND CAPPINGS

Copings and cappings are designed to protect the tops of walls and finish them neatly. In practice they will do so only if skill and care is constantly exercised by the bricklayer.

THEIR PURPOSE

On brick parapet, freestanding and retaining walls they greatly reduce the quantity of water penetrating and possibly saturating the brickwork below.

Both bricks and mortar that are frequently saturated should be specified as frost resistant in order to avoid frost attack. Saturated mortar may also crumble from sulfate attack if it is not correctly specified, gauged and mixed.

The appearance of brickwork that becomes saturated may be spoilt temporarily by efflorescence or virtually permanently by lime-leaching or algae. In the long term badly made joints between coping units may cause local saturation and staining.

DEFINTIONS

* Copings
  Copings have an overhang with a throat to shed run-off rainwater clear of the brickwork immediately below (fig 5.1).
* Cappings
  Cappings are usually flush with the wall below. They may have an overhang but no throat (fig 5.2).

MATERIALS

Copings, cappings and mortar jointing will usually suffer severe saturation and freezing and should have been specified for their durability under such conditions. Copings and cappings may be made from:

* Precast concrete
* Natural or artificial stone
* Bricks – either standard or special shaped bricks
* Brick and tile creasing
* Terracotta
* Slate
* Metal or plastic – not usually fixed by bricklayers.

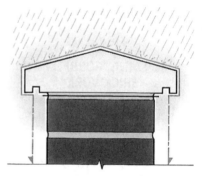

Figure 5.1. *A coping.*

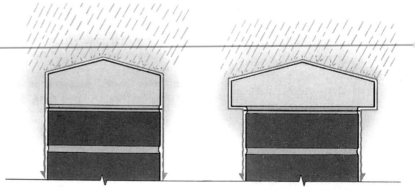

Figure 5.2. *Cappings.*

PRECAST CONCRETE AND STONE COPINGS

- **Damp-proof courses**
  Copings should be bedded on a damp-proof course to prevent moisture reaching the brickwork through the inevitable fine cracks which will develop between the mortar and the coping units. The DPC should preferably project about 10 mm from the face of the wall.
  The DPC and coping should be bedded in one operation on to fresh mortar to achieve the maximum bond with the wall below.

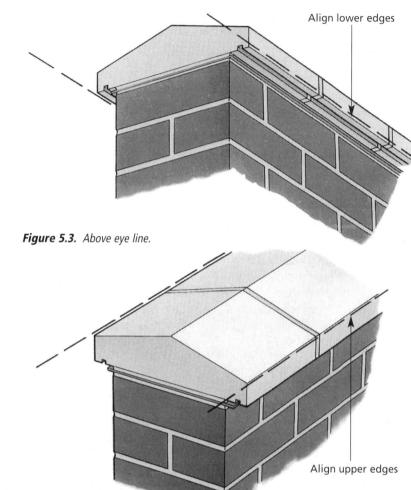

Figure 5.3. *Above eye line.*

- **Method of alignment**
  If the coping is above eye level, as on a parapet or high boundary wall, the *lower* edge should be 'lined-in' *(fig 5.3)*. On a low wall the *upper* edge should be aligned *(fig 5.4)*.

For a narrow coping only a front line is necessary *(fig 5.5)*.

Figure 5.4. *Below eye line.*

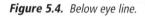

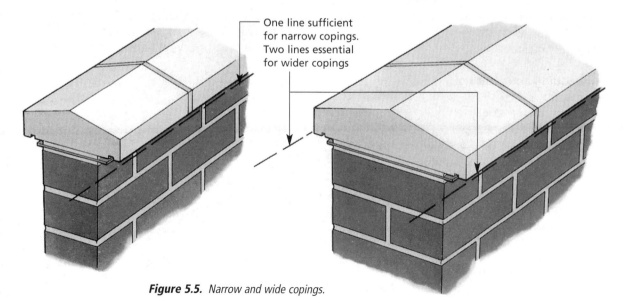

Figure 5.5. *Narrow and wide copings.*

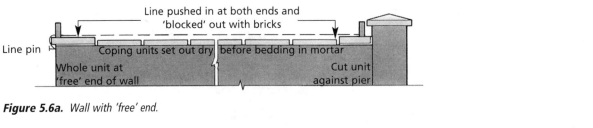

Figure 5.6a. Wall with 'free' end.

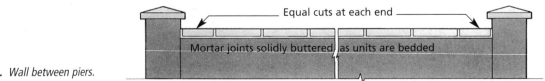

Figure 5.6b. Wall between piers.

- **Fitting, cutting and bedding**
  Determine how many coping units and cuts will be required. Allow for 6–10 mm wide joints between units. Cut units should not be positioned at the 'free' end of a wall *(fig 5.6a)*. If the coping is between piers, place equal cut lengths at each end *(fig 5.6b)*.

For neat straight cuts and to minimise breakages use a disc cutter. Bed a coping unit at each end and pull a line between. Before bedding the rest of the units set them out 'dry' to establish a suitable and consistent gauge.

- **Bedding and jointing mortar**
  Gauge mortar strictly as instructed because the proportions will have been specified to produce a durable mortar. The mix will usually be 1:¼:3, cement:lime:sand (a designation (i) mortar).
  The cement, sand and any pigments should each be obtained from a single source throughout the job so that the mortar will not vary in

colour and make the finished work look patchy.

- **Movement joints**
  Care should be taken to position movement joints in copings and cappings strictly as specified. They should coincide with movement joints in the wall below. Movement joints must be kept free of mortar droppings and debris and filled only with the specified jointing material – never with mortar as it is not compressible.

- **Throats**
  The throat should be continued through the jointing material otherwise staining or deterioration of

the brickwork below may occur *(fig 5.7)*.

- **Cramps**
  Brush out any dust from the slots and dampen them before bedding the non-ferrous cramps using mortar of the same mix as the bedding mortar.
  Such cramps are often used on sloping parapet gable walls *(fig 5.8)*. They must never be fixed across a movement joint.

Brick copings and cappings
Brick copings may be built from standard half-round or saddleback units or from proprietary units.

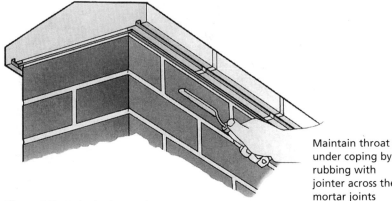

Figure 5.7. Maintain throats.

Maintain throat under coping by rubbing with jointer across the mortar joints between units

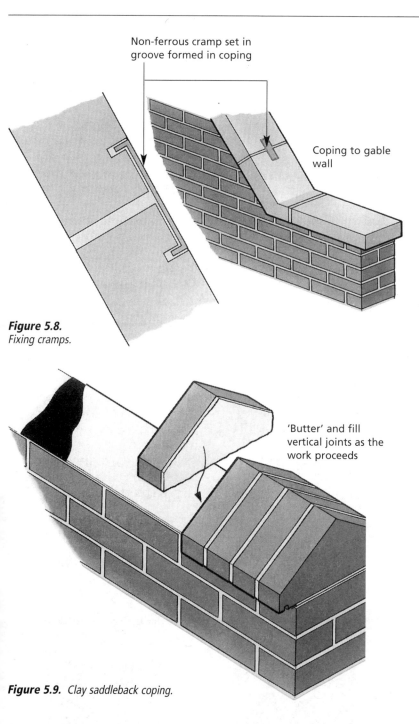

Figure 5.8.
Fixing cramps.

Non-ferrous cramp set in
groove formed in coping

Coping to gable
wall

'Butter' and fill
vertical joints as the
work proceeds

Figure 5.9. *Clay saddleback coping.*

If the dimension of the brick coping units, along the length of the wall, is greater than the standard 65 mm (some are 215 mm) they should be built as described above for precast concrete and stone copings.

Cappings, if flush with the wall below, are simpler to 'line-in'.

All the comments concerning DPCs, Bedding and jointing mortars, Movement joints and Throats made under the heading 'Precast concrete and stone copings' apply when building brick copings.

- **Damp-proof course**
  After building a brick-on-edge capping it is advisable to lay a course of bricks dry on top while the mortar sets in order to improve its bond with the bricks and the DPC *(fig 5.10).*

Some consider that the top of the wall will be more stable if the DPC is placed under the course below the capping.

This consideration does not apply where the DPC is part of a proprietary interlocking capping system.

The DPC should be left slightly projecting or cut flush as specified. It should never be cut back from the face and pointed over, as this often leads to spalling of the mortar or even the bricks.

- **Alignment**
  Lines (•) should be positioned to prevent 'tipping' and 'waving'. Where possible use one line to maintain alignment of a 'preferred' face or 'give and take' if the brick dimensions are not

When bedding 65 mm thick units, ensure that the vertical joints are fully buttered and filled as the work proceeds *(fig 5.9).*

Preferably begin by building 'starter blocks' on the ground or other suitable surface and allow to harden. Then bed one at each end of the wall and pull lines through. Position two lines, one back and one front, to prevent the units from 'tipping' especially when using half-round coping bricks.

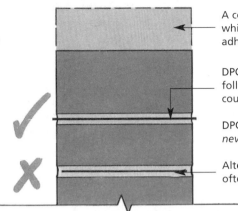

A course of 'dry' bricks placed while the mortar sets may help adhesion of top course

DPC bedded on fresh mortar followed quickly by the next course

DPC left projecting or cut flush, *never* cut back from surface

Alternative lower position for DPC often preferred

Figure 5.10. *A standard brick-on-edge capping.*

consistent. Use a line on the other side to prevent 'tipping' *(fig 5.11).*

- **Long walls**
  The lining-in of long stretches of copings and cappings necessitates the 'eyeing-in' or 'sighting' of intermediate points, between which the units are run-in, to check that the line is tight enough. Tingle plates prevent long lines from sagging and must

also be 'sighted' or 'eyed-in' *(fig 5.12).*

- **Brick and tile creasing course**
  This consists of a brick-on-edge course set on two courses of tiles known as creasing courses.
  The tiles may be virtually flat with no nibs. Tiles with continuous nibs which are carefully built to form a continuous throat may enable

brick and tile creasing courses to function as a coping if a flexible DPC is also included in the assembly. The DPC should be sandwiched in mortar below the tile creasing courses. The creasing courses alone will not prevent the downward percolation of water into the brickwork below.

Lay two courses of creasing tiles in mortar to 'half bond' taking care to 'wipe on' mortar at each butt joint between tiles. The first course should be lined-in along the lower edge and the second course along the upper edge. Creasing tiles are seldom regular in shape and some 'give and take' may be necessary *(fig 15.13).*

Select the required number of bricks for the capping course and check for consistency of size, discarding any with excessive variation from the average length and width. The guidance given for the selection of bricks for a soldier arch is appropriate *(Section 5.5).*

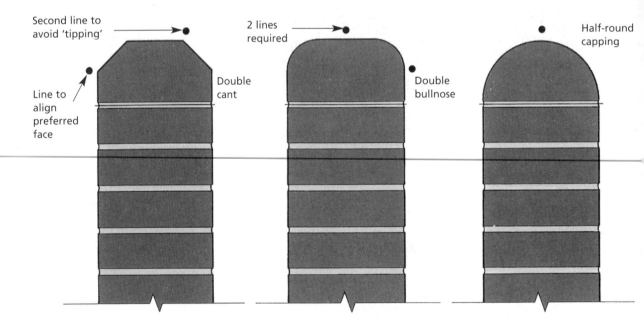

Second line to avoid 'tipping'

Line to align preferred face

Double cant

2 lines required

Double bullnose

Half-round capping

Figure 5.11. *Special shapes – brick-on-edge cappings.*

The brick-on-edge course should be bedded, the vertical joints fully buttered as the work proceeds and any frogs solidly filled.

It is best to begin by preparing end blocks of three bricks unless special stop ends are specified.

When the blocks have set, bed one at each end to which a line may be fixed (fig 5.14).

Set out the gauge to avoid cutting bricks. This is preferably done by tightening rather than opening joints.

The brick to be laid should be buttered and pushed against the existing brickwork (fig 5.15a).

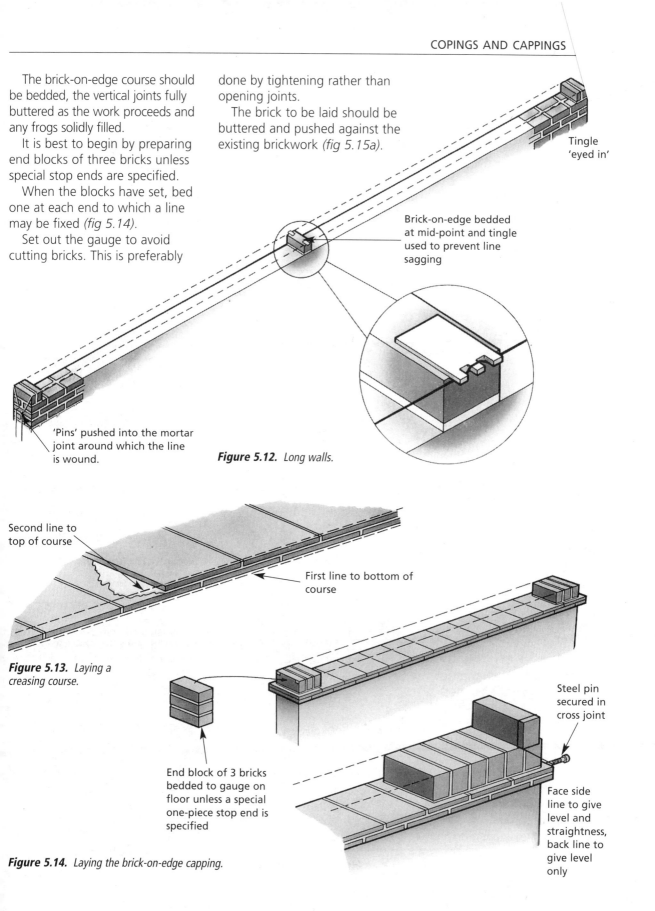

Tingle 'eyed in'

Brick-on-edge bedded at mid-point and tingle used to prevent line sagging

'Pins' pushed into the mortar joint around which the line is wound.

Figure 5.12. Long walls.

Second line to top of course

First line to bottom of course

Figure 5.13. Laying a creasing course.

Steel pin secured in cross joint

End block of 3 bricks bedded to gauge on floor unless a special one-piece stop end is specified

Face side line to give level and straightness, back line to give level only

Figure 5.14. Laying the brick-on-edge capping.

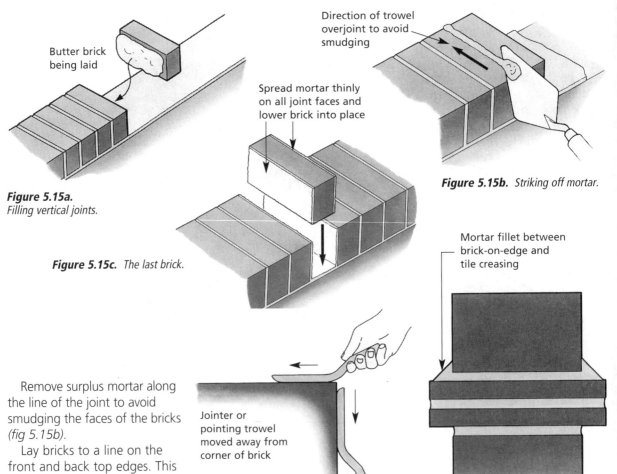

Butter brick
being laid

Figure 5.15a.
Filling vertical joints.

Direction of trowel
overjoint to avoid
smudging

Figure 5.15b. *Striking off mortar.*

Spread mortar thinly
on all joint faces and
lower brick into place

Figure 5.15c. *The last brick.*

Mortar fillet between
brick-on-edge and
tile creasing

Jointer or
pointing trowel
moved away from
corner of brick

Figure 5.16. *Finishing joints.*

Figure 5.17. *Final fillet.*

Remove surplus mortar along the line of the joint to avoid smudging the faces of the bricks (fig 5.15b).

Lay bricks to a line on the front and back top edges. This keeps the whole course in line and prevents 'waving'. Check the gauge frequently to avoid a sudden change in joint width.

To place the last brick, butter the bricks already laid together with a small amount on both sides of the brick to be laid. Gently push the brick into place ensuring that mortar does not build up to smudge the brick faces (fig 5.15c).

To finish the joints between the bricks on edge, pull the jointer away from the corner to leave a clean line and compact the mortar at the arris (fig 5.16). Finally, run a mortar fillet between the tile creasing and the brick-on-edge capping if required (fig 5.17).

NOTE: The Brick Development Association does not consider that tile creasing will continue to be an effective DPC in resisting the downward flow of water over a long period of time. A continuous DPC below the tile creasing, although not common practice, is strongly recommended to ensure its continuing effectiveness. Brick-on-edge and tile creasing is not considered be completely effective as a coping unless the tiles have a continuous nib from which a continuous throat is formed.

KEY POINTS

- Bed DPCs on fresh flat mortar either projecting or cut flush with mortar joints.
- Work out gauge to prevent unnecessary and unsightly cutting.
- Align 'high' copings along lower edge and 'low' copings along upper edge.
- Consider the 'sight line' before fixing the lines and pins.
- Fill all joints solidly as the work proceeds.
- Remove excess mortar by running the trowel along the line of the joint to prevent smudging.
- Keep checking gauge to maintain an even joint size.
- Maintain throat across joints between coping units.

5.2 CAVITY PARAPET WALLS

Too often parapet walls are designed or built so badly that rain penetrates or the brickwork cracks or suffers frost or sulfate attack. Putting the faults right is usually expensive and disruptive for the occupants.

This section explains the principles of good design and construction by reference to the most common form of parapet, i.e. those projecting above flat or pitched roofs, although they are also used elsewhere including bridges.

Figure 5.18. High parapet wall to accessible flat roof.

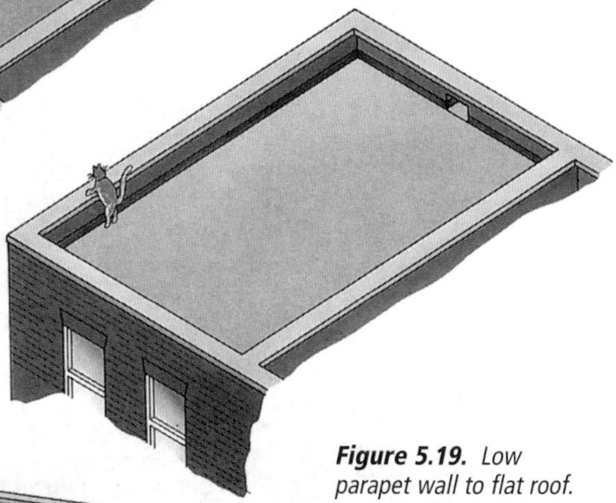

Figure 5.19. Low parapet wall to flat roof.

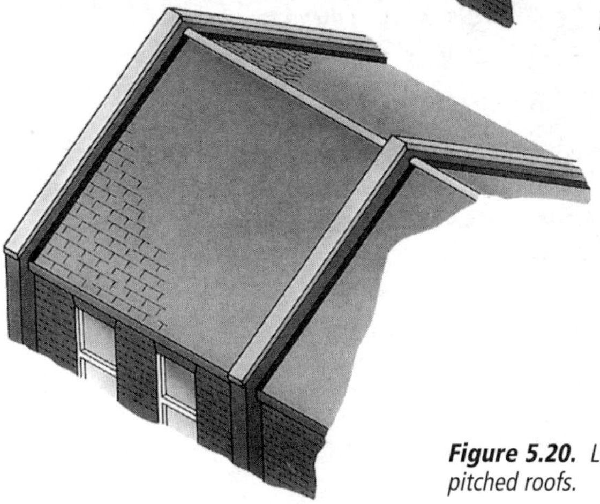

Figure 5.20. Low parapet walls to pitched roofs.

High parapets provide a safety barrier at the edge of roofs to which people have ready access. They should be designed by structural engineers to resist strong winds and people leaning against them *(fig 5.18)*.

Low parapets to flat and pitched roofs may be only the minimum height for practical construction *(figs 5.19 & 5.20)*.

Both faces of a brick parapet wall must be built to line, level, gauge and plumb and left clean and jointed. But, in addition they must be able to resist the very severe exposure of the three surfaces to wind-driven rain and extremes of temperature.

The following checklist of operations will help bricklayers avoid the most common faults. The typical examples illustrated are of facing brickwork cavity construction, as solid parapets are not recommended in these circumstances, being more liable to rain penetration. Rendered parapets require further consideration which is beyond the scope of this section.

BUILDING OPERATIONS

DPC trays

1 Build DPC trays into the inner leaf not less than 150 mm above the finished roof *(figs 5.21a, b & c)*. Use only specified DPCs and in particular **never** polythene or other low adhesion DPCs which may cause a plane of weakness and allow rain penetration between the DPC and the

mortar. *(see Section 4.3 'Damp-proof courses')*

2 Step cavity trays up or down to the outer leaf as instructed, by not less than 150 mm. If the cavity contains thermal insulation the tray **must** step down to the outside leaf to protect the top of the insulation which should be level with the bottom of the tray *(figs 5.21a & b).*
  Trays should also be sloped down to the outer leaf in highly exposed conditions as otherwise water may track across the underside of the cavity tray after penetrating under the DPC *(fig 5.21c).* Failure to achieve good adhesion by bedding the DPC on fresh mortar will increase this risk.

3 Stepping trays down to the outer leaf increases the risk of staining below weep holes although the risk is not great and is invariably preferable to rain penetration. Proprietary fibrous filter plugs are sometimes positioned in weep holes with the intention of reducing staining by filtering solid particles from the drained water. The risk of staining can also be reduced by clearing mortar droppings from the cavity tray before closing the cavity.

4 Bed DPC trays on fresh mortar followed by the next course as soon as possible to achieve good adhesion between mortar and DPCs. DPCs should project 5 mm, or be flush as instructed. **Never** position DPCs so that the edges are covered with mortar as this can cause spalling of the mortar and brick edges as the DPC compresses under load. *(see Section 4.3 'Damp-proof courses')*

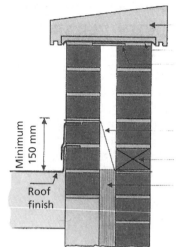

PC concrete coping

DPC bedded on fresh mortar
Rigid support

Clean mortar droppings from tray

Weep holes at minimum of 1 m centres

Carry insulation up to underside of cavity tray

Minimum 150 mm

Roof finish

Figure 5.21a.

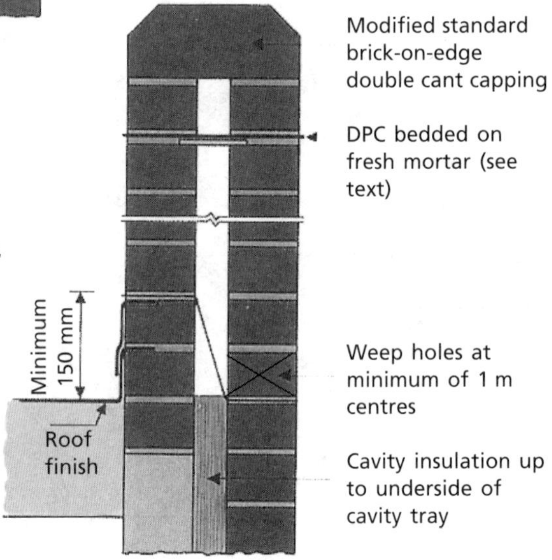

Figure 5.21b.

Modified standard brick-on-edge double cant capping

DPC bedded on fresh mortar (see text)

Weep holes at minimum of 1 m centres

Cavity insulation up to underside of cavity tray

Minimum 150 mm

Roof finish

Figure 5.21c.

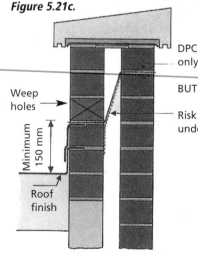

Weep holes

Minimum 150 mm

Roof finish

DPC stepped down to inner leaf only if no insulation in cavity

BUT

Risk of rain penetration along underside of tray (see text)

Figure 5.21. *Typical alternative parapet details.*

5 Cavity trays in a parapet wall will often have to drain away large quantities of water, and if there are any poorly sealed joints some water will leak into the insulation and the inner leaf.

Joints, in long runs of cavity trays ('running laps') or with purpose-made return units, must be lapped by at least 100 mm and sealed *(fig 5.22)*. Use either a liquid adhesive or an adhesive tape as recommended by the DPC manufacturer.

Unless permanent proprietary joint-supports, usually of polypropylene, are specified, improvise a plywood off-cut to support the lap temporarily while the two portions are pressed together.

The use of preformed cavity tray units at corners and changes of level is simpler than cutting and forming sheet DPC on site.

Metal flashings at abutments of roof and parapet walls

The purpose of metal flashings is to weather the junction between DPCs and roof finishes.

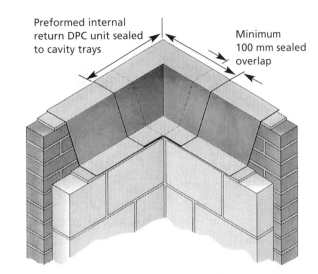

Preformed internal return DPC unit sealed to cavity trays

Minimum 100 mm sealed overlap

Figure 5.22. *Sealing DPC trays and units where lapped.*

Unfortunately poor design and/or building practice often result in rain penetration.

6 Flashings must be in the same joint as DPCs and under them *(fig 5.23a)*. If the flashing is over the DPC *(fig 5.23b)* or in another joint *(fig 5.23c)*, rainwater may penetrate.

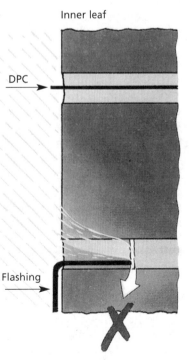

Inner leaf

DPC

Figure 5.23. *Right and wrong ways of positioning DPCs and flashings.*

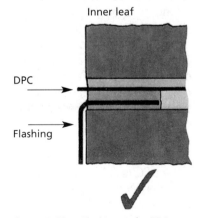

Figure 5.23a. *Flashing under DPC.*

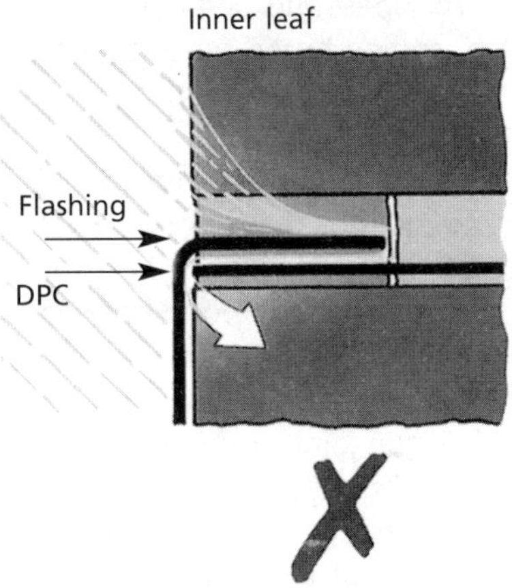

Figure 5.23b. *Flashing over DPC – possible rain penetration.*

Figure 5.23c. *Flashing in course below DPC – possible rain penetration.*

NOTE: If lead DPcs are in contact with free lime from mortar for prolonged penods in very wet conditions, they should be protected from corrosion by a thick coat of bitumen paint on both sides. However, the Lead Sheet Association does not consider such protection necessary for lead flashings tucked only 40–50 mm into mortar joints, since in such conditions, relatively near the drying surface, carbonation of free lime is rapid and there is said to be no risk of corrosion.[1]

THERE ARE TWO BASIC METHODS OF INSTALLING FLASHINGS

(1) Flashings are fixed into unfilled joints or chases left by the bricklayer

7 After bedding the DPC tray on fresh mortar, the bricklayers should rake the green mortar from **below** the DPC, to a depth of 25 mm to leave room for the flashing to be inserted and fixed at a later date.

8 BUT it is very difficult to fill such a thin joint with sealant (mortar would be ineffective) and wedging is virtually impossible (fig 5.24a). In practice the joint below the DPC should be at least 8 mm thick. Such a thick joint is normally unacceptable in facework but this can be overcome by cutting rebates in the bricks on a masonry bench saw (fig 5.24b). Cutting a chase into the finished brickwork, with an angle grinder, is possible but difficult.

9 When the flashing is fixed it should be wedged every 450 mm, and a suitable backing material inserted so as to leave the correct depth for pointing with a sealant which is

Figure 5.24. *Fixing a flashing – practical considerations.*

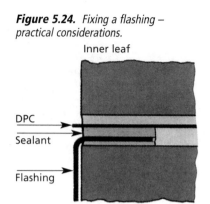

Inner leaf

DPC
Sealant
Flashing

Figure 5.24a. *Difficult in practice.*

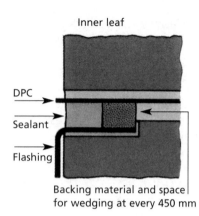

Inner leaf

DPC
Sealant
Flashing

Backing material and space for wedging at every 450 mm

Figure 5.24b. *Bricks rebated to provide a wider joint and a more practicable operation.*

Inner leaf

DPC cut back when chopping out mortar to receive flashing

Figure 5.25. *DPC removed in error leads to rain penetration.*

compatible with the DPC material. Information on sealants and backings should be obtained from the manufacturer of the particular sealant to be used.

Two-part and one-part polysulfide-based sealants should be to the relevant British Standards.[2][3]

Care should be taken when fixing wedges, as brickwork less than 600 mm high may be lifted, particularly if the bond between the mortar and DPC is weak.

10 If the joint is not raked out by the bricklayer, the trade fixing the flashing is likely to destroy the edge of the DPC when chopping out the hardened mortar (fig 5.25).

(2) Flashings built-in by the bricklayer

11 For this method, the Lead Sheet Association[4] recommends that flashings are single welted and formed so that at least 50 mm rests on the brickwork (fig 5.26). The DPC tray is bedded on fresh mortar followed immediately by the next course of bricks. The welt anchors the flashing in the mortar and avoids the use of lead wedges and the risk of lifting brickwork as mentioned in 9 above. Experience seems to suggest that if the DPC is well bedded, the joint between it and a lead flashing will remain weathertight.

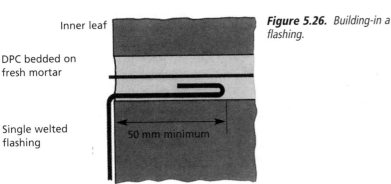

Inner leaf

DPC bedded on fresh mortar

Single welted flashing

50 mm minimum

Figure 5.26. *Building-in a flashing.*

Weep holes

12 Form weep holes in vertical cross joints at a maximum of 1 m intervals in the course laid on the lower side of the tray. Leave weep holes clear of debris and insert filter plugs if specified.

Copings and cappings

(see Section 5.1 for definitions)

13 Clean mortar droppings and debris from cavity trays to prevent debris being washed through the weep holes by rainwater penetrating the cavities and causing stains.

14 Bed the DPC on fresh mortar for maximum adhesion and on a rigid support across the cavity to prevent the DPC sagging into the cavity which will channel rainwater percolating through the coping or capping, into the cavity *(fig 5.27)*. Bed the coping or capping immediately after the DPC to load the mortar joint and get the maximum adhesion between DPC, mortar and coping. *(see Section 5.1 'Copings and cappings')*

In *Figure 5.21b* a DPC is shown one course below the brick-on-edge capping, rather than immediately below it, in order to provide more weight and better adhesion and stability

at the joint containing the DPC. This DPC may not be specified in a very low parapet if it is only one or two courses above the cavity tray.

Proprietary systems of special interlocking bricks and incorporating a DPC have been developed by several brick manufacturers in order to provide greater weight and stability. They should be built strictly in accordance with the manufacturers' recommendations.

Mortar joints

15 Fill all bed and cross joints solidly with mortar and finish with weathered or bucket handle joints to maximise the rain resistance of both sides of the parapet wall.

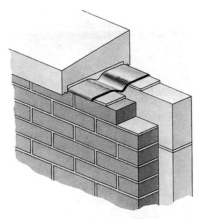

Figure 5.27. *Unsupported DPC sags and forms path for water passing through cracked mortar joints to enter cavity.*

Movement joints

16 Because parapet walls are highly exposed to extremes of temperature from both sun and wind and are not restrained by the weight of any structure above, movement joints will be required at closer intervals than in the main walls and must be carefully built *(see Section 4.5 'Vertical movement joints')*

The right bricks

17 Clay bricks used for copings, cappings and for parapet walls which are topped with cappings rather than copings must be frost resistant (F). Calcium silicate bricks in cappings should be at least class 4. *(see Section 6.4 'Durability of brickwork' tables 6.1 & 6.3)*

The right mortar

In most conditions only mortar mixes of designations (i) and (ii) are suitable. Designation (iii) is acceptable only when there is a low risk of saturation, e.g. in a low parapet which has a coping.

Sulfate-resisting cement is recommended, in order to resist sulfate attack where there is a high risk of saturation **and** where clay bricks with Normal (N) rather than Low (L) levels of soluble salts as defined in BS 3921[5] are used. *(see also Section 6.4 'Durability of brickwork' table 6.1)*

Some specifiers may require the use of additives such as styrene butadiene to increase the water resistance of the mortar joint.

For more detailed guidance and recommendations see section 6.4 'Durability of Brickwork' and in particular table 6.5 – Masonry condition F 'Unrendered parapets'.

References

(1) 'The Lead Sheet Manual' 1990. The Lead Sheet Association.

(2) BS 4254 'Specification for two-part polysulphide-based sealants'.

(3) BS 5215 'One-part gun-grade polysulphide-based sealants'.

(4) Figures 5.23, 5.24, 5.25 and 5.26 are based on diagrams in 'The Lead Sheet Manual 1990 – volume 1 Lead Sheet Flashings' with acknowledgements to the Lead Sheet Association.

(5) BS 3921:1985 'British Standard specification for clay bricks'.

KEY POINTS

- Check type of bricks are as specified or recommended.
- Check mortar mixes are as specified or recommended.
- Check if extra vertical movement joints are required.
- Bed all DPCs on fresh mortar.
- Project DPCs at face or build-in flush.
- Seal all laps in DPCs with adhesive or tape.
- Generally bed copings and cappings on DPCs supported over cavity.

5.3   CURVED ARCHES

The arch, developed early in the history of masonry construction, elegantly transfers loads from above wall openings to abutments each side. The Romans built arched viaducts and aqueducts, and in Victorian Britain arches proliferated for spectacular civil engineering structures as well as simple houses *(fig 5.28)*.

Today, most arches are built into the outer leaf of cavity walls and are 'self-supporting'. But the increasing use of structural masonry in recent years means that some arches will be 'structural', supporting roof, floor and wall loads. In either case the eye is readily attracted to the arch form, and bricklayers need to take particular care with the appearance. Additional engineering requirements must be met when building structural arches.

Although many brick manufacturers supply sets of voussoirs ready to build into arches, bricklayers who can set out arches and cut bricks to shape will be well respected for having a sound, broadly based-knowledge of their craft As such they will be in demand for the more advanced, high quality facing and structural brickwork which will be increasingly required.

This section covers the basic skills required when building curved arches. The building of soldier arches is described in section 5.5.

Figure 5.28a. *A 19th century brick viaduct.*

Figure 5.28b. *Victorian granary – Bristol.*

Figure 5.28c. *Victorian houses – Lichfield.*

Figure 5.29. *A rough two-ring semi-circular arch.*

ROUGH ARCHES

Rough arches are built from standard, parallel-sided bricks with wedge-shaped joints *(fig 5.29)*. The larger the radius the less the joints taper.

Their use is normally confined to semi-circular or segmental arches and is probably most acceptable when rugged soft-mud or stock type bricks are used. Smooth, even-coloured bricks seldom look well with tapered joints.

Although the bricks need no preparation it may be necessary to discard a few which vary unacceptably in size from the remainder.

AXED ARCHES

Traditionally, axed arches are built from voussoirs fair cut on site from standard bricks using a lump hammer, bolster and scutch. The joints are parallel sided and nominally 10 mm wide

(fig 5.30). This method is satisfactory only with relatively soft bricks. A masonry bench saw is required to cut hard bricks. *(see Section 2.5 'Cutting bricks')*

The basic geometry necessary to draw part of an axed arch full size, in order to make and traverse a template for cutting arch bricks, is described in text books such as Hodge's 'Brickwork for apprentices'.[1]

Some manufacturers will, on request, prepare computer-aided drawings (CAD) giving the sizes of voussoirs for particular arches, and some will cut them to shape before delivery.

GAUGED ARCHES

Traditionally, gauged arches were built from bricks known as 'rubbers' made from a fine, red-burning clay, blended with a high percentage of fine sand. They were soft enough to cut and rub to shape and size on site. Very fine joints of 1 mm or less were achieved using lime putty *(fig 5.31)*.

A few manufacturers make traditional 'rubbers' but they are expensive and are used mainly for high quality restoration work. Some manufacturers offer bricks of a similar appearance, mechanically cut and sometimes

rubbed to shape. Depending on the type of brick, joints 2–3 mm thick, using lime putty made with silver sand, may be achieved. But with cement:sand mortars, joints from 6–10 mm will be more typical.

The craft of building gauged brickwork is fully described in Gerard Lynch's book 'Gauged Brickwork'.[2]

TEMPORARY SUPPORTS

All brick arches need temporary support during construction *(figs 5.32a & b)*. Both the types of arch centres illustrated can be used many times.

Place the arch centres on folding wedges and timber props each side of openings. The folding wedges:

- give fine adjustment when levelling the arch centre.
- enable centres or turning pieces to be gently lowered ('easing' and 'striking') when the mortar joints have hardened i.e. after 28 days for 'structural' arches and 14 days for 'self-supporting' arches.

Today, proprietary, permanent, metal arch support systems or reusable polystyrene are often supplied to site for use instead of traditional timber centring *(fig 5.32c)*.

PREPARATION

- Raise the brickwork abutments both sides of an arch to the level of the springing line and place the arch centre or turning piece in position.

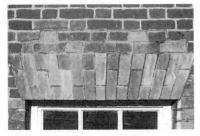

Figure 5.30. *An axed cambered arch.*

Figure 5.31. *A gauged segmental arch.*

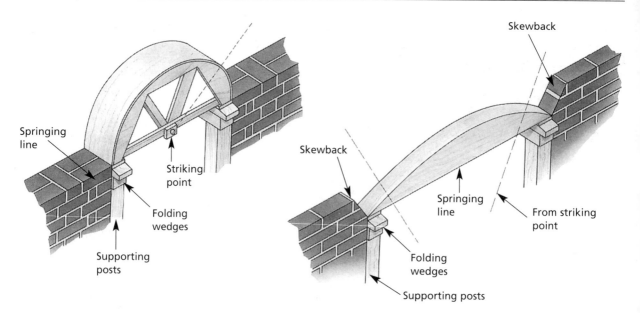

Figure 5.32a. *Traditional open timber framework arch centre for a semi-circular arch.*

Figure 5.32b. *Solid timber turning piece for a segmental arch in a half-brick leaf.*

Figure 5.32c.
Polystyrene centre in use.

- For segmental arches, raise further courses of walling to form the skewback bearings *(fig 5.32b)*, making use of a 'gun template' to obtain the correct angle. *(see Section 5.9 'Tumbling-in courses', figures 5.95 & 5.98 showing a gun template in use)*
- Next plumb up from the striking point to locate the midpoint of the key brick position. With rough ring arches having two or more separate rings, the second

and subsequent even-numbered rings must have a central joint, not a key brick.

CONSTRUCTING A ROUGH RINGED ARCH
- Mark the width of the key brick in pencil on the arch centre and mark a joint allowance each side *(fig 5.33)*.
- Using a flexible steel tape, measure down the curve of the intrados from the key brick space or centre joint to

the springing line and divide into a **whole** number of equal sized brick spaces including a joint allowance which should be as small as possible. Otherwise the wedge-shape joints become too wide at the extrados.

In practice the minimum joint, using cement:sand mortars, is probably 6 mm.

- Set the first arch brick at the springing line on bedding mortar built up slightly thicker at the extrados in order to form a wedge-shaped joint *(fig 5.34)*.
- Continue placing wedge-shaped bedding mortar on the arch brick previously laid in order that the bricks follow the curve. Build up both sides alternately so that the centre is loaded evenly.
- Set arch bricks accurately to the pencil mark spacings, and

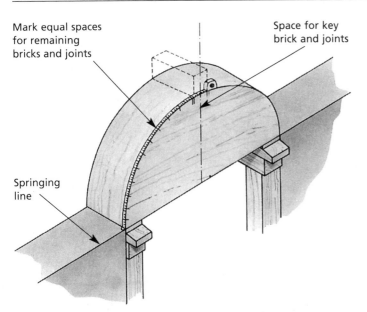

Mark equal spaces for remaining bricks and joints

Space for key brick and joints

Springing line

Figure 5.33. *Marking out a rough ring arch on an arch centre.*

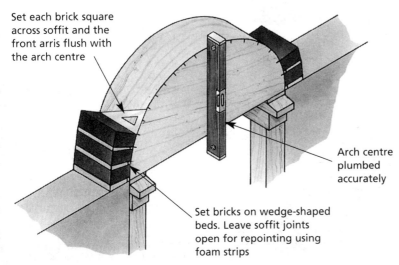

Set each brick square across soffit and the front arris flush with the arch centre

Arch centre plumbed accurately

Set bricks on wedge-shaped beds. Leave soffit joints open for repointing using foam strips

Figure 5.34. *Setting bricks accurately on plumbed centre.*

square across the soffit *(fig 5.34)*.

- Set bricks dry on the arch centre by bedding back and leaving a joint space for repointing at a later date. Insert foam or wood strips at the soffit *(fig 5.34)*.
- Set the key brick in place ensuring that mortar is solidly packed into these last two

top joints at the crown of the arch.

CONSTRUCTING AN AXED SEGMENTAL ARCH

The following applies whether purpose-made moulded arch bricks are supplied by a manufacturer or voussoirs have been cut on site by a bricklayer

working from a full-size drawing from which a cutting template has been made. See brickwork textbooks for detailed procedure of setting out an axed arch, making and traversing a cutting template, e.g. Hodge.[1]

Set up a temporary arch support as before and mark on top, in pencil, the width of the key brick *(fig 5.35)*. The width will have been determined on an arch setting out drawing prepared either by the architect, brick manufacturer or by the bricklayer on site.

- Mark a joint space, shown on the arch drawing, at each side of this key brick space.
- Using dividers, mark out in pencil the required number of spaces each side of the key brick corresponding to the number of voussoirs on the arch drawing. Each space allows for a voussoir and one joint.
- Bed the first voussoir at the springing of the arch and check joint alignment using a piece of string fixed to the striking point of the arch *(fig 5.36)*. Continue bedding and checking voussoirs. Work alternately on each side so as to load the turning piece (arch centre) evenly.
- Keep soffit joints clear of mortar for later repointing as previously described.
- Ensure voussoirs are square on the soffit and follow pencil markings precisely.
- Constantly check face plane alignment with straight edge or line and pins as the arch progresses *(fig 5.37)*.
- When it becomes impractical to apply mortar to the

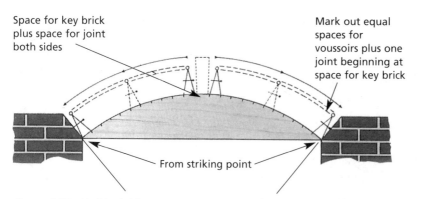

Space for key brick plus space for joint both sides

Mark out equal spaces for voussoirs plus one joint beginning at space for key brick

From striking point

Figure 5.35. *Marking brick spaces on a turning piece for a segmental arch.*

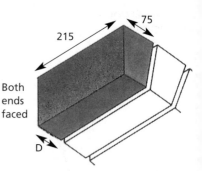

215   75

Both ends faced

D

Figure 5.38a. *Tapered header type AR.1 to BS 4729.*

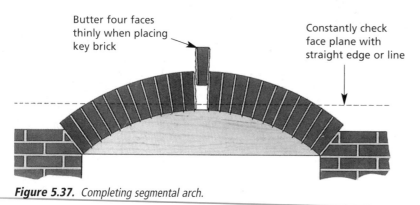

Work alternately from each side

Check line of voussoirs with string from striking point

Figure 5.36. *Building axed segmental arch.*

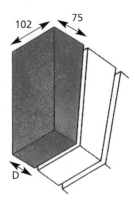

102   75

D

Figure 5.38b. *Tapered stretcher type AR.2 to BS 4729.*

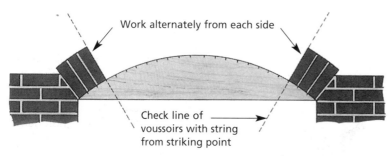

Butter four faces thinly when placing key brick

Constantly check face plane with straight edge or line

Figure 5.37. *Completing segmental arch.*

facilities to determine the exact sizes and numbers required. Enquiries should have been made with manufacturers at an early stage in the design process if purpose-made voussoirs rather than manufacturer's standard sets are required.

previously laid brick, 'butter' each voussoir evenly across the bedding surface before setting.

Complete key brick jointing in a similar manner to that described in section 5.1 'Copings and cappings' *figure 5.15.*

*Figure 9 inside the rear cover shows a good example of a large gauged arch.

MANUFACTURED ARCH SETS

Many brick manufacturers now offer to supply, to the client's requirements, sets of purpose-made arch bricks. They may be sized to be built with nominal 10 mm joints or, using lime putty, 2–3 mm joints to simulate traditional gauged brickwork.* Most manufacturers have computer-aided design (CAD)

BRITISH STANDARD ARCH BRICKS

BS 4729[3] 'Dimensions of Bricks of Special Shapes and Sizes' specifies tapered headers (type AR.1) and tapered stretchers (type AR.2) *(figs 5.38a & b)*.

The alternative dimensions for the smaller end of the wedges (D) are to suit semi-circular arches of four different spans which are multiples of whole numbers of bricks. All these dimensions are set out in Table 8 of the Standard.

DAMP-PROOF TRAYS OVER ARCHES

All features which bridge a cavity, with the exception of wall ties, must be protected from water running down inside the cavity by an effective cavity tray which will collect the water and discharge it to the outside. *(see Sections 3.3 'External cavity walls'; 4.3 'Damp-proof courses'; 6.7 'Rain resistance of cavity walls')*

Cavity trays over curved arches are difficult to form on site where the slightest imperfection can lead to rain penetration and expensive remedies. Experience suggests that these vital accessories are most satisfactory when they are purpose-made by specialists.

KEY POINTS

- Use an accurate centre or turning piece.
- Provide strong, properly braced temporary support that will not collapse under the weight of bricks and mortar during construction.
- Raise centre or turning piece up to the required springing line of the finished arch.
- Always ensure that an arch centre is perfectly plumb on face before setting the arch bricks.
- Always plumb up from the striking point to locate the centre of the key brick.

- Pre-plan and pencil mark positions for all the arch bricks around the centre before starting work.
- Prevent mortar from getting under the arch bricks and onto the temporary support when building an arch.
- Keep soffit joints clear of mortar to simplify pointing later.
- When building an arch from purpose-made bricks or with axed voussoirs always use a length of string fixed to the striking point to check alignment.

References

(1) Hodge, J. C. (1971) 'Brickwork for apprentices'. London: Edward Arnold
(2) Lynch, G. C. J. (1990) 'Gauged brickwork – a technical handbook'. Aldershot: Gower Technical.
(3) BS 4729:1990 'Dimensions of bricks of special shapes and sizes'.

5.4 CIRCULAR BULL'S-EYES

Great care is required to construct truly circular bull's-eye openings in brick walls.

Vertical and horizontal diameters of finished openings must be absolutely equal for a satisfactory appearance and for a circular window frame to fit the space.

A bull's-eye consists of two identical semi-circular arches, but the method of constructing the lower half is quite different from turning the upper half.

CRAFT TERMS

Terms in italics are illustrated in fig 5.39.

CONSTRUCTION

Most bull's-eyes are of axed brickwork formed by wedge-shaped bricks called *voussoirs* with parallel joints between. Wedge-shaped joints between uncut bricks in a rough-ring bull's eye are generally considered unsightly.

It is possible for a skilled bricklayer to set out an axed bull's-eye, make a cutting template and produce voussoirs by hand using a hammer, bolster and comb hammer. See section 5.3 'Curved arches'. Bench-mounted masonry saws would be required to cut both hard and perforated bricks cleanly.

However, most brickmakers can readily supply purpose-made tapered arch bricks provided they are ordered in advance of site requirements. The majority of manufacturers have computer-aided design (CAD) facilities to determine the sizes of arch bricks.

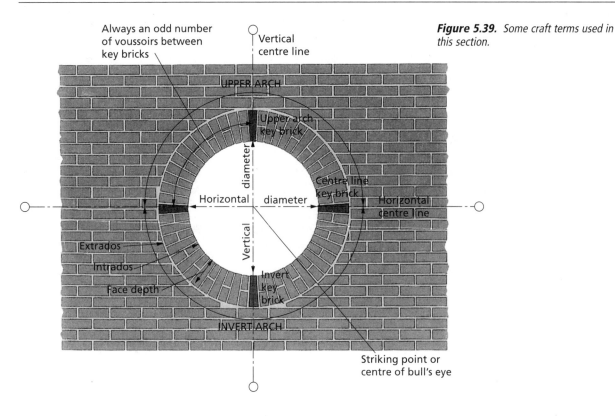

Always an odd number of voussoirs between key bricks

Vertical centre line

UPPER ARCH

Upper arch key brick

Vertical diameter

Horizontal diameter

Centre line key brick

Horizontal centre line

Extrados

Intrados

Face depth

Vertical diameter

Invert key brick

INVERT ARCH

Striking point or centre of bull's eye

Figure 5.39. *Some craft terms used in this section.*

SETTING-OUT

A bull's-eye is set out with key bricks on both the *horizontal and vertical centre lines* which means that if the *face depth* is bonded each quadrant must have an odd number of voussoirs between key bricks to avoid joints in adjacent voussoirs coinciding *(fig 5.39)*. If special shape bricks, type AR.2.1 to BS 4729[1], were used to build this bull's-eye it would require 50 rather than 56 voussoirs as the extrados (outer curved surface) face dimension of the standard arch brick is 75 mm, being based on the former culvert or 9-inch radial end arch brick.

BUILDING SEQUENCE

Building a circular bull's-eye begins with the lower half or *invert arch* using a trammel. No temporary supports are used as they are for the *upper arch* which is built in a similar way to a normal semi-circular arch.

Brickwork up to the horizontal centre line

The brick wall is built up to the horizontal centre line of the bull's-eye, racking back from the extrados of the invert arch *(fig 5.40)*.

The wall should be sufficiently high to support the timber beam from which the trammel will swing. The brickwork must be accurately lined through and plumb on face so that the entire circumference of the bull's-eye will align with the completed wall.

The trammel

A 100 × 75 mm timber, or similar, is bedded temporarily on the brickwork in lime:sand mortar, spanning the opening, and the *striking point or centre* of the bull's-eye is marked. Weighting it with bricks ensures that it will not move *(fig 5.40)*.

The trammel is cut from a suitable batten, say 25 × 6 mm. The length from the centre nail to the point is the outer radius of the bull's-eye **plus** a mortar joint. The trammel is nailed to the centre point mark on the timber beam and should swing freely *(fig 5.40)*.

The invert

The invert is carefully cut to the required semi-circular shape. Each brick is temporarily placed in position using wood spacers normally 10 mm thick, to represent bed and cross joints *(fig 5.41)*. The curve is marked

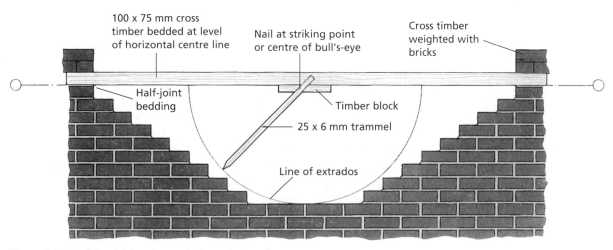

Figure 5.40. *Building brickwork up to horizontal centre line.*

with a pencil held at the point of the trammel. The bricks are cut to shape with a bolster chisel or masonry bench saw as appropriate to the type of brick, and finally dressed to the curve with a comb hammer. Some manufacturers of hard bricks will cut the end of bricks to the shape of the extrados when 'green' leaving the bricklayers to make a vertical cut to fit the bond pattern. They are then permanently bedded in position, checking with the trammel *(fig 5.41)* as each cut brick is set until the invert has been completed. *(see Sections 2.5 'Cutting bricks'; 6.9 'Bricklaying tools and equipment')*

As the bricks at the bottom of the invert usually require long tapering cuts the use of machine cutting may be the only practicable method with many types of bricks. A final rub with a carborundum stone will improve the curved shape as each of these cut bricks is formed and ensure that the lower half of the bull's-eye is neatly bedded.

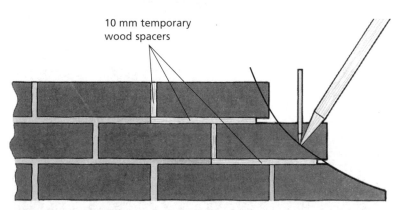

Figure 5.41. *Marking the curve before cutting.*

Voussoirs in the invert arch
Without disturbing the timber beam, carefully remove the trammel and drill an additional hole 225 mm closer to the trammel point. Re-fix the trammel to the beam carefully. This can now be used to set the voussoirs by swinging the invert of the bull's-eye. A string line is fixed to the centre nail to check that each voussoir radiates from the centre point. Bed the bottom key brick, checking with trammel and string line. The positions of the voussoirs are pencil-marked

around the invert starting from the key brick using dividers and working to left and right.

The voussoirs are bedded from the *invert key brick* alternately either side until the invert arch is complete up to the *horizontal centre line*. Constant checks of the face plane must be made using a level or straight edge *(fig 5.42)*.

Voussoirs in the upper arch
The upper half is built in a similar way to any semi-circular arch. *(see Section 5.3 'Curved*

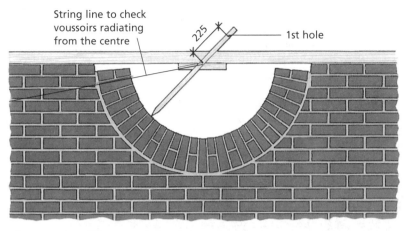

Figure 5.42. Building voussoirs into invert arch.

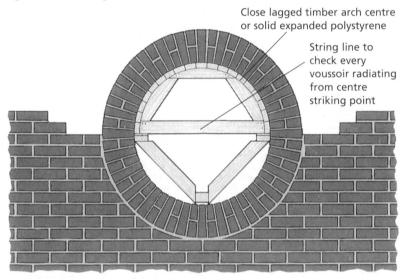

Figure 5.43. Building voussoirs into upper arch on arch centre.

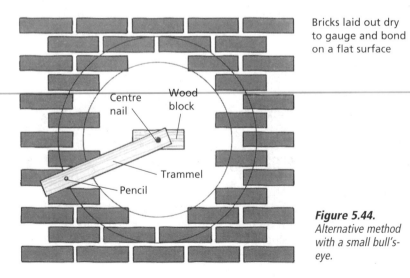

Figure 5.44. Alternative method with a small bull's-eye.

arches.) An arch centre is set up and wedged in position and a string line fixed to the centre point nail *(fig 5.43)*. Using a steel tape check that the vertical diameter is exactly the same as the horizontal diameter.

The voussoir positions are set out with dividers and marked with a pencil around the arch centre. Voussoirs are bedded using the string line to ensure that they radiate from the centre point of the bull's-eye and the face-plane alignment is checked with a level or other straight edge. Alignment of the arch face with the wall face is aided with a line and pins set up along the face of the wall.

Before each voussoir is bedded a nominal 10 × 10 mm strip of polystyrene or tapered wood is positioned to prevent a build-up of mortar on the soffit by forming recessed soffit joints for subsequent repointing with a matching mortar *(see Section 5.3)*.

Brickwork above the horizontal centre line

The brickwork surrounding the arch is marked and cut with the same care as that at the invert. Parallel **nominal** 10 mm joints of even width should be maintained between all voussoirs and around the circumference of the bull's-eye.

After an absolute minimum of seven days and when the mortar has set hard and sufficient brickwork built above to stabilise the arch, the timber centre is removed by easing the wedges and removing the struts and the soffit is pointed to match the facework mortar.

MARKING OUT CUT BRICKS SURROUNDING A SMALL BULL'S-EYE – AN ALTERNATIVE METHOD

A small diameter bull's-eye of say 600 mm may be marked and cut using a circular template of plywood or hardboard instead of a trammel. The bricks are laid on a flat surface with spacing for joints and the template placed on top. The curve is then marked on the bricks with a pencil, cutting and setting the invert as before *(fig 5.44)*.

<table>
<tr><td colspan="2">

KEY POINTS

</td></tr>
<tr><td>

■ Ensure bull's-eye is a true circle.
■ Weight timber beam firmly so that it does not move during use.
■ Space voussoirs evenly.

</td><td>

■ Use string line from striking point to ensure voussoirs radiate correctly.
■ Use line and pins to keep face of bull's-eye lineable with the wall face.

</td></tr>
</table>

Reference
(1) BS 4729:1990 'Specification for dimensions of bricks of special shapes and sizes'.

5.5   SOLDIER ARCHES

A soldier arch is a course of bricks set on end and having a flat soffit *(fig 5.45)*. It is not a structural arch form and so a permanent independent loadbearing support is provided.

Soldier arches in the brick outer leaves of cavity walls are usually supported by a proprietary steel lintel, the type depending on the loading, cavity widths and the thickness of the inner leaf.

Alternatively, they may be supported by a simple steel angle which will show beneath the arch or may be suspended from an angle or brackets above by steel reinforcement. Reinforced soldier arches are built on temporary timber supports (head trees) which are removed when the mortar has cured. *(see Sections 4.6 'Reinforced and post-tensioned brickwork'; 4.7 'Brickwork on metal support systems')*

Figure 5.45. A typical soldier arch.

CONSTRUCTION
This section describes the basic bricklaying operations for constructing a soldier arch supported by a steel box lintel.

Selection of bricks
Because soldier arches are conspicuous features, the bricks from which they are to be constructed should be carefully selected to ensure that:

• they do not vary greatly in length. Experience suggests that the difference in length of the shortest and longest bricks should be no more than 3 mm.
• unacceptably bowed, twisted or chipped bricks are rejected.

Raising reveals
After the lintel has been bedded in position, raise three courses above soffit level as a vertical extension of the reveals *(fig 5.46)*.

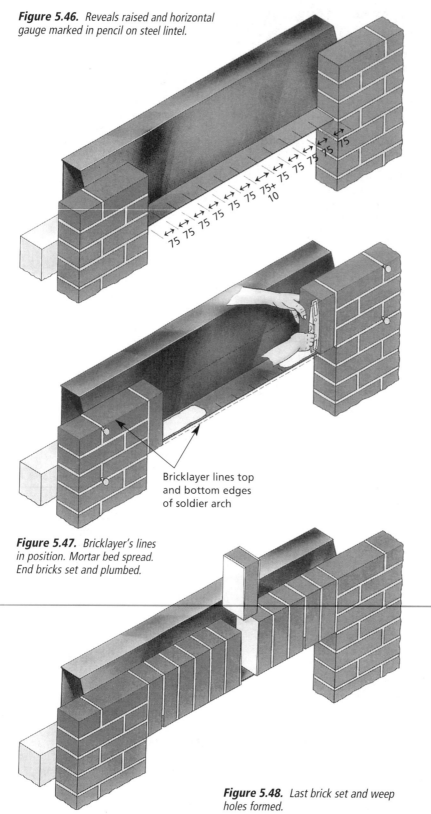

Figure 5.46. *Reveals raised and horizontal gauge marked in pencil on steel lintel.*

Bricklayer lines top and bottom edges of soldier arch

Figure 5.47. *Bricklayer's lines in position. Mortar bed spread. End bricks set and plumbed.*

Figure 5.48. *Last brick set and weep holes formed.*

These will provide support for the first and subsequent arch bricks as they are set in position from each end.

Setting out

Mark in pencil on the steel lintel, a horizontal gauge for the bricks in the soldier course at intervals of 75 mm from each end. If the opening is to standard gauge, there will be an extra 10 mm gap left between the last two marks. If not, evenly adjust the marks so that a whole number of bricks will fit *(fig 5.46)*.

Fix two bricklayers' lines to align and level the top and bottom of the arch *(fig 5.47)*.

Spread a mortar bed along the toe of the lintel.

Butter the first brick solidly, taking great care not to smear the face, slightly furrow the mortar to assist joint compaction when tapping the brick plumb, and set in position at one end. Without moving your hand from the brick check its verticality with a boat level *(fig 5.47)*. Because the eye is drawn to a soldier arch every brick must be perfectly plumb.

Support each brick with one hand while tapping it plumb, so that mortar suction is given a chance to hold the brick steady.

To place the last brick, thinly butter the bricks already laid and both sides of the brick to be laid. Gently push the brick into place ensuring that mortar does not build up to smudge the brick faces *(fig 5.48)*.

Form weep holes by removing mortar from the lower third of vertical cross joints at no more than 1 m centres. Generally, there should be no less than two over any opening. If required,

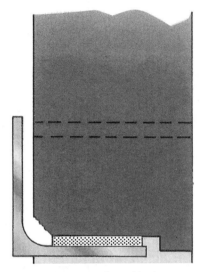

Figure 5.49. Use of pistol bricks on a steel angle support.

insert or build-in proprietary preformed plastic ventilator/weep holes or fibrous filters at the face of the joint.

Some other points

Pistol bricks may be supplied to site for bedding on lintels and steel angles in order to avoid a thick bed joint *(fig 5.49)*. When setting bricks on steel angles it may be necessary to chip off the rear arris to allow them to bed correctly.

When building on temporary timber supports (head trees), bricks are laid dry, not bedded in mortar. First, fix suitable lengths of foam strips or wood on the temporary support to keep the bottom of the soffit joints clear of mortar so that they can be pointed when the support is removed. *(see Section 4.7 'Brickwork on metal support systems')*

Foam strips are better than wood for preventing leakage of wet mortar.

5.6 DECORATIVE BRICKWORK*

Simple brickwork in half or quarter bond often provides all the colour, texture and pattern that is required. But sometimes it is enriched by the use of different coloured bricks; modelling the surface with projecting and recessed bricks; bedding bricks vertically or at an angle to the normal horizontal bedding pattern or using bricks of special shapes.

The basic techniques described in this section can be adapted by skilled bricklayers to build a variety of decorative brickwork.

POLYCHROMATIC BRICKWORK

Polychromatic brickwork is built from bricks of different colours in decorative features ranging from simple band courses of contrasting colour to complex patterns or murals. Coloured mortars may be used to match or contrast with the bricks.

COLOURED MORTARS

Trial panels, to determine the required mortar mix, should show at least 100 brick faces, be in accordance with appendix F of BS 3921[1] and reflect the form of the decorative brickwork to be

built. The mortar joints should be raked and pointed with different mixes which must be allowed to dry before inspection. *(see Section 1.1 'Reference and sample panels')*

Repeatedly producing mortar to a consistent colour on site can be difficult and time consuming. Changing mixes, either deliberately or through inaccurate measurement of the quantity of pigments, sand, lime and cement or changing the type or source of sand and even the cement, can result in colour changes in the mortar and patchy brickwork.

*See inside covers for examples of decorative brickwork.

Consistent jointing techniques are necessary. For example, if the mortar joints are too wet

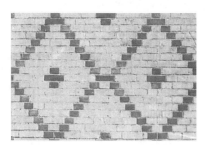

Figure 5.50a. *Diaper pattern in quarter bond.*

Figure 5.50b. *Flamboyant Victorian brickwork based on diaper pattern.*

when finished by ironing or tooling, the fines are brought to the surface so that the joint dries a lighter colour. *(See also Sections 2.8 'Pointing and jointing'; 6.6 'Appearance' especially under heading 14 'Maintain a consistent jointing technique')*

Suppliers of ready-mixed lime:sand (coarse stuff) provide colour charts and consistent mixes. If coarse stuff is used it is still necessary to measure accurately the amount of cement added on site.

DIAPER WORK

Many different diaper patterns can be created by coloured, projecting or recessed bricks *(fig 5.50).*

If different coloured mortars are specified it is generally preferable to use one mortar throughout and rake the appropriate joints to a depth of 12 to 15 mm and point with coloured mortar. Using two different coloured mortars is difficult, requiring two trowels

and two spot boards.

When building-in projecting bricks take particular care with line, level and plumb *(fig 5.51).* See also 'Projecting bricks' on page 143.

BAND COURSES

Bricks supplied for band courses *(fig 5.52)* may have a different average length from other bricks in the wall. Generally, set out all courses to the co-ordinating dimensions of the bricks (225 mm for standard bricks in stretcher bond) by tightening or opening vertical joints as necessary and plumbing every fourth or fifth perpend. *(see Section 2.4 'Vertical perpends')*

Soldier courses

Soldier courses consist of bricks set on end showing stretcher faces. The bricks should be selected, either by the brickmaker or on site, to have a close tolerance about the mean length, typically ± 1.5 mm.

For greatest accuracy lay to a line. But a line secured to freshly

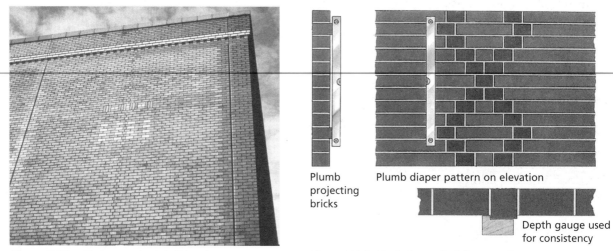

Plumb projecting bricks

Plumb diaper pattern on elevation

Depth gauge used for consistency

Figure 5.50c. *Modern version in half bond.*

Figure 5.51. *Vertical control of diaper work.*

Figure 5.52a. *Bold use of band courses in contrasting colours. (See also inside back cover.)*

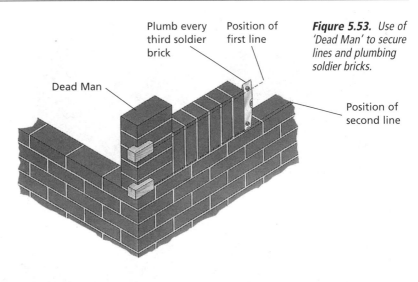

Figure 5.53. *Use of 'Dead Man' to secure lines and plumbing soldier bricks.*

Plumb every third soldier brick

Position of first line

Dead Man

Position of second line

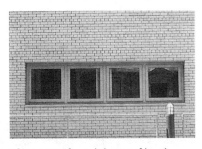

Figure 5.52b. *Subtle use of band courses in snapped headers.*

laid soldier bricks may pull them out of plumb when pulled taut. To avoid this:

- Preferably build 'Dead Men' at the ends or corners to secure the line and build the soldier course between *(fig 5.53)*, then replace each 'Dead Man' with three soldier bricks.
- Alternatively, construct the corners in the traditional manner *(see Section 2.3 'Line, level and plumb')* but do not run out on the level more than nine or ten soldier

bricks as they are very difficult to keep to line level and plumb because of the small bed area.

The eye is readily attracted to soldier courses so take extra care to set the bricks plumb. As a general rule plumb every third soldier with a boat level *(fig 5.53)*. Butter the soldier brick to be laid, **not** the face of the last brick laid. In addition ensure that the **top** of the course lines through. For comments on laying the last brick at the centre, see section 5.1 'Copings and cappings' *figure 5.15* and section 5.5 'Soldier arches'.

Alternative methods of returning soldier courses are shown in *fig 5.54*. Note that *figs 5.54a & b* do not allow, on adjacent return walls, the same relationship between the vertical cross joints in the soldier courses and those in the course below. Architects sometimes require vertical cross joints in a soldier course to coincide with those in the course below, even though bricklayers are usually taught to avoid this.

Bricklayers may be required to build-in special ties to restrain

the smaller return bricks such as that in *fig 5.54a* as they may be vulnerable to displacement by lateral movement of walls at the corners.

HERRINGBONE BOND

There are six main types of herringbone bond *(fig 5.55)*. They are usually built as a panel surrounded by normal half or quarter bonded brickwork, or between window openings. Setting out and building single vertical herringbone bond is described in some detail below as an example. Brief comments are made on other herringbone bonds.

Building single vertical herringbone bond

Build the surrounding brickwork accurately to the correct height, level, plumb and gauge *(fig 5.56)*. Ensure that the tops of the reveals are at the same level. Keep the opening width constant for the whole height. This is best done with a pinch rod rather than relying only on keeping the reveals plumb.

a. Standard special SD.1

b. Standard special SD.2

c. Standard special BD1.3

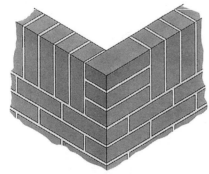

d. Standard bricks stack bonded

e. Bonded corner

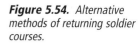

Figure 5.54. *Alternative methods of returning soldier courses.*

a. Vertical herringbone.

b. Double vertical herringbone.

c. Horizontal herringbone.

d. Double horizontal herringbone.

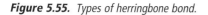

e. Diagonal herringbone.

f. Double diagonal herringbone.

Figure 5.55. *Types of herringbone bond.*

Setting-out and cutting

Transfer a full-size outline of the opening on to a suitable board *(fig 5.57)*. Deduct 20 mm (the total thickness of the 10 mm joints at each reveal) from the length and 10 mm from the height to allow for the first bed joint. Do not deduct 10 mm for the joint at the top of the panel as the top of the herringbone panel will be level with the top bricks of the reveals.

On the board, mark horizontal and vertical centre lines and lines at 45° to them through the centre point. This gives the position of the first bricks to be placed in the setting-out process *(fig 5.57)*.

Starting with two bricks *(fig 5.57)*, place all the bricks dry and accurately on the board *(fig 5.58)*. Any inaccuracy at this stage will be reflected in the built panel.

Next, transfer the outline of the board to the bottom and two sides of the dry bricks and cut the bricks to shape.

Bricks for decorative work are best cut on a masonry bench

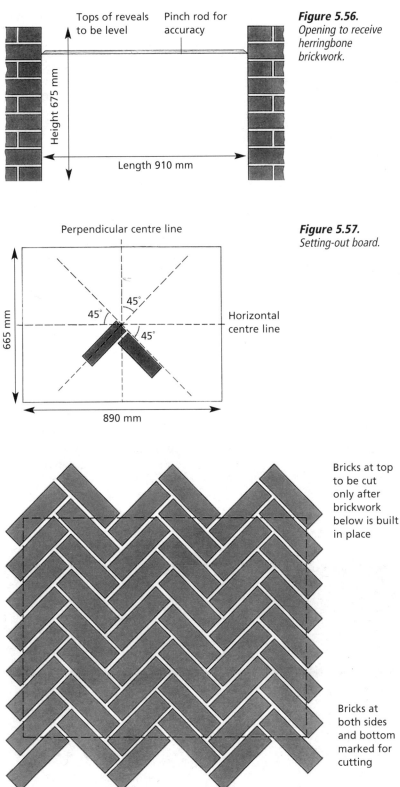

Figure 5.56. *Opening to receive herringbone brickwork.*

Tops of reveals to be level

Pinch rod for accuracy

Height 675 mm

Length 910 mm

Figure 5.57. *Setting-out board.*

Perpendicular centre line

Horizontal centre line

45° 45° 45°

665 mm

890 mm

Bricks at top to be cut only after brickwork below is built in place

Bricks at both sides and bottom marked for cutting

Figure 5.58. *Bricks set-out dry and marked for cutting.*

saw. If it is necessary to cut an acute angle by hand, use a sharp bolster and first cut the brick square and then trim the angle to minimise the risk of the 'point' breaking off. When cutting by hand, rest the brick on a bed of sand or a pad of old carpet to avoid breaking the bricks. *(see Section 2.5 'Cutting bricks')*

Construction
After cutting the bricks to shape, place a gauge rod (cut to the width of the opening) along the bottom of the dry panel and mark the position of the cut bricks. Use the gauge rod as a template to position the bricks in the first course.

The laying of herringbone bond must be controlled at an angle of 45°. Traditionally a 'boat level' and 45° set square were used, but 'boat levels' are now available with adjustable vials *(fig 5.59)*. As the vials must be accurate the levels should be obtained from a reputable manufacturer. Run a level line to control the top of each course.

Lay the last row temporarily and mark them with a chalk line, level with the top of the reveals, and cut accurately.

Ensure mortar is of the right consistency to avoid staining from wet mortars and so that the bricks can be gently 'rubbed' into position. Tapping or knocking down bricks is likely to disturb the work below far more in angled brickwork than in horizontal courses.

Double vertical herringbone bond
In order to 'centre' the pattern, set the centre bricks in relation to the centre lines as shown *(fig 5.60)*.

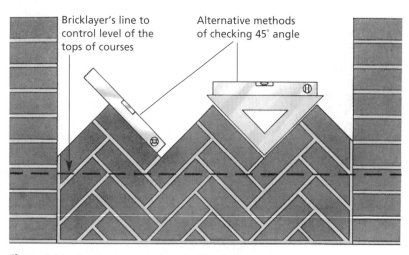

Figure 5.59. *Constructing a single vertical herringbone panel.*

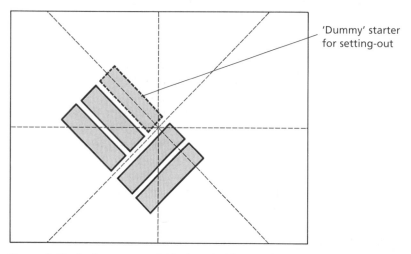

Figure 5.60. *Setting-out centre bricks for a double vertical herringbone panel.*

Horizontal and double horizontal herringbone bond

Setting-out and construction are as for the vertical bonds except that the direction of the bonds is horizontal rather than vertical *(figs 5.55c & d).*

Diagonal and double diagonal herringbone bond

These bonds have approximately only a third as many cuts as the horizontal and vertical forms. Because the pattern aligns with the main brick courses it can be built at the same time, thus avoiding the lengthy setting-out process *(figs 5.55e & f).*

BASKET WEAVE BOND

Basket weave is usually formed with three stretchers stack bonded and three 'soldiers' laid adjacent *(fig 5.61a).*

Unlike horizontal and vertical herringbone bond, basket weave does not need to be set out dry, and no cutting is required, but the opening must be kept to brick sizes. Because level and plumb are critical, a boat level should be used on each brick as

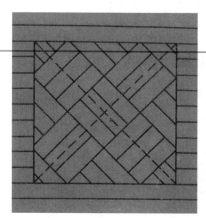

a. Basket weave   b. Diagonal basket weave

c. Alternative diagonal basket weave

Figure 5.61. *Basket weave bonds.*

well as checking for level and gauge by the line.

Diagonal basket weave bond

Being at 45°, diagonal basket weave bond *(fig 5.61b)* does require setting-out and special construction as described for vertical and horizontal herringbone bond.

Setting-out starts at the centre so that the complete panel is symmetrical. The centre of the middle brick can be positioned at the centre of a square as a guide for setting-out with the 45° lines passing directly through the centre *(fig 5.61b)*. Alternatively, the main continuous joints can be set out to form an 'X' through the centre of a square panel *(fig 5.61c)*.

INTERLACING BOND

Interlacing bond, sometimes referred to as 'pierced panelling', has ⅓ brick cuts left open and has been used for garden walls *(fig 5.62a)*.

The panel consists of whole bricks laid vertically and horizontally, with ⅔ bricks to achieve the interlacing effect.

During construction lay the ⅓ cut bricks in sand. Remove them when the mortar has set hard.

Alternatively the ⅓ cuts can be bricks of contrasting colour laid in mortar. The panel follows brick courses and can be laid as work proceeds, but ensure the vertical pattern is kept plumb as this will be the 'line of sight'.

Diagonal interlacing bond

Interlacing bond *(fig 5.62b)* cannot be built as a pierced panel. As with all panels with 45° bonds, much cutting is required.

a. Interlacing bond

b. Diagonal interlacing bond

Figure 5.62. *Interlacing bonds.*

Figure 5.63a. *A dog tooth course of blue bricks in a wall of London stocks.*

Figure 5.63b. *A recessed dog tooth course with a dentil course below – by trainees at West Kent College.*

DOG TOOTHING

There are two basic forms of dog toothing. In both, two faces are set at 45° to the line of the wall face *(figs 5.63a & b)*. In one, each arris between the two faces projects 72 mm from the face of the wall and in the other they are flush with the wall, forming a recessed feature.

Building projecting dog toothing

Lay the first brick 'dry' on the wall, mark and cut it accurately for use as a template for the remaining bricks *(fig 5.64a)*.

A line should be positioned at the top of the arris to control the projection *(fig 5.64b)*.

Place a 1.2 m spirit level along the underside of the string course to ensure that the 'line of sight' is maintained. Level the course from back to front regularly *(figs 5.64c & d)*.

Set-out the corner details first and then set-out the bond over the whole length. Open or tighten the vertical joints as necessary because it is not possible to introduce cuts. Line the cut surface of each dog tooth brick with the inside face of the external leaf to maintain

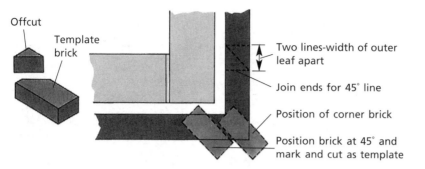

Offcut

Template brick

Two lines-width of outer leaf apart

Join ends for 45° line

Position of corner brick

Position brick at 45° and mark and cut as template

Figure 5.64a. *Marking 45° line on top of outer leaf – positioning and cutting template brick.*

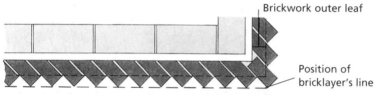

Brickwork outer leaf

Position of bricklayer's line

Figure 5.64b. *Plan of dog tooth course.*

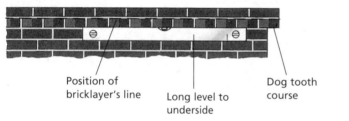

Position of bricklayer's line

Long level to underside

Dog tooth course

Figure 5.64c. *Elevation of dog tooth course.*

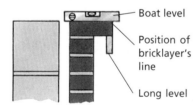

Boat level

Position of bricklayer's line

Long level

Figure 5.64d. *Vertical section through dog tooth course.*

Figure 5.64. *Constructing a dog tooth course.*

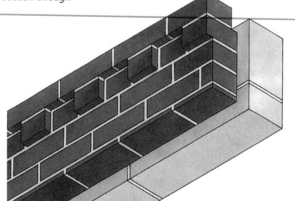

Figure 5.65. *A dentil course.*

Figure 5.66. *A mural depicting a steel worker – in Sheffield. (See inside back cover for colour illustration.)*

the 45° angle. But check it with a bevel.

Building recessed dog toothing

Proceed in a similar manner as described for projecting dog toothing but position the line on the face line of the wall. As this makes contact with the arrises only, check the level of the dog toothing every three or four bricks with a spirit level.

See the paragraph headed 'Projecting bricks' on page 143.

DENTIL COURSES

A dentil course consists of a regular pattern of projecting headers cut to project 28 mm. In cavity walls, the cut is flush with the inner face of the outer leaf *(fig 5.65)*. The feature is finished with a course of stretchers laid centrally over the headers either flush or projecting a further 28 mm.

The projecting headers should be placed to avoid a straight joint with the course below. See the paragraph headed 'Projecting bricks' below.

KEY POINTS

Because decorative brickwork is eye catching, pay particular attention to the following:

- **Bricks with close dimensional tolerances are advisable for most decorative brickwork. If specified, ensure that only such bricks are used.**
- **General accuracy of plumb line and gauge.**
- **Plumb of soldier bricks.**
- **Accuracy of projections and overhangs – they cast shadows.**
- **Consistency of mortar to allow rubbing down of diagonal brickwork.**
- **Accurate control of bricks at 45° angles.**
- **Neat and consistent jointing and pointing.**
- **Neat and accurate cutting of bricks.**
- **Cut dog tooth bricks accurately so as not to protrude into the cavity.**
- **Protect finished work, especially projections, as work proceeds.**

MURALS

Bricklayers are required to work to detailed drawings for the position and colour of each brick in polychromatic murals such as that in *fig 5.66*.

PROJECTING BRICKS

When bricks are being specified for decorative brickwork that includes projecting bricks, consideration should be given to the following:

- Perforations and frogs may be exposed to view and the weather, in which case solid bricks will generally be advisable.
- Exposed bed surfaces of standard bricks will usually be different in appearance from the normal face but can be matched as a special requirement as type BD.1.3 to BS 4729[(2)].
- For some types of bricks, opposite stretcher and header surfaces differ in appearance. Check whether purpose-made bricks with two faced stretchers have been provided, e.g. for the return of a dog tooth course.
- Exposed projecting bricks should be specified frost resistant.

References

(1) BS 3921:1985 'Specification for clay bricks'.
(2) BS 4729:1990 'Dimensions of bricks of special shapes and sizes'.

5.7 CURVED BRICKWORK

When building straight walls to line, level, gauge and plumb, the most important items in a bricklayer's tool kit are 'line and pins'. But as these cannot be used for curved walls, bricklayers adopt other methods as described in this section.

ALIGNMENT WITH TEMPLATES

Alignment is most commonly checked with a curved template made from a flat board about 1200 mm long *(fig 5.67)*.

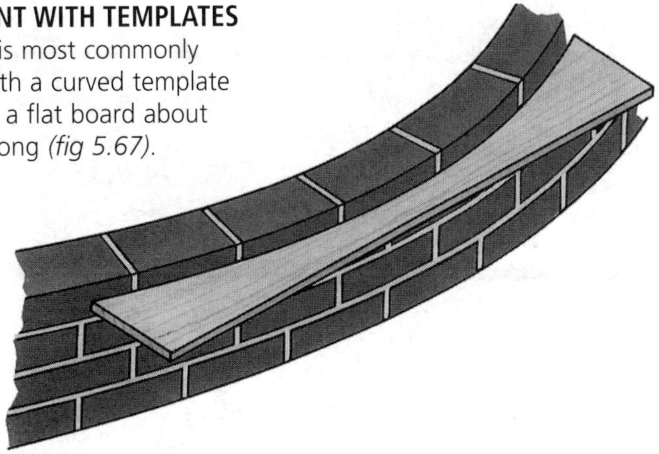

Figure 5.67. *Timber template approx 1.2 m to check alignment of curved course.*

SETTING-OUT

Straight walls are set-out using string lines between foundation profiles.

To set out **curved** walls, find from the drawings, and locate on site, the position of the **striking point (A)** and *the radius of curvature (B)* *(fig 5.68)*.

The builder will dig curved foundation trenches and place concrete on which the bricklayers can set-out the curved walling.

TRAMMEL AND PLUMB

Using a trammel (radius rod) and spirit plumb rule, plumb down into the trench and mark points in a thin mortar screed *(fig 5.69)*. Use a template to join the points and mark a complete curve locating the face side of the whole wall *(fig 5.70)*.

BONDING

Walls that are curved on plan can be built from straight bricks by forming wedge-shaped mortar cross joints. A stretcher bonded convex curved half-brick leaf (102 mm thick) can be built to a diameter as little as 3 m without the need to cut the back corners and to give an acceptable cross joint on the face.

For thicker walls of various face bonds, bricklayers should 'strike' the radius of curvature on a flat surface and lay out the bricks 'dry' around the curved line. This will indicate the amount of cutting required, widths of cross joints on the face and whether Flemish or Header bond is possible where a small radius is required.

For the best appearance, curved walls are built with curved bricks known as 'radials' *(fig 5.71)*. Six standard radial headers and six standard radial stretchers, specified in BS 4729[1], give convex walls with six ideal outer radii from 450 mm to 5.4 m. These are summarised in table 5.1.

The use of radial bricks should allow cross joints to have parallel sides rather than be wedge shaped. Further detailed information on radial bricks and their use is given in tables in BS 4729[1] and the Brick Development Association's publication, 'The design of curved brickwork'[2]. The latter also includes information on the use of straight bricks in curved walls.

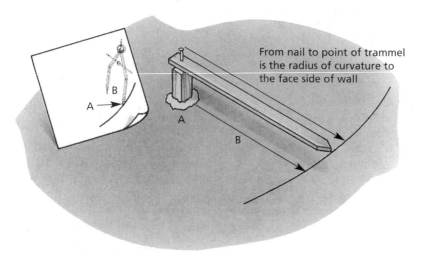

From nail to point of trammel is the radius of curvature to the face side of wall

Figure 5.68. *Timber peg at striking point A, and trammel to set out radius of curvature B as found from drawings.*

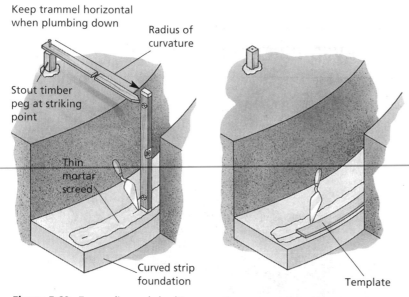

Keep trammel horizontal when plumbing down

Radius of curvature

Stout timber peg at striking point

Thin mortar screed

Curved strip foundation

Figure 5.69. *Trammeling and plumbing down position of curve.*

Template

Figure 5.70. *Completing line of curve with template.*

TABLE 5.1	Standard radial bricks – to BS 4729		
Type no.	Dimension 'D'	Ideal outer radius	No. of bricks in quadrant
RD.1.1	52	450	6
RD.1.2	70	675	9
RD.1.3	80	900	12
RD.1.4	89	1350	18
RD.1.5	97	2250	30
RD.1.6	103	5400	72
RD.2.1	172	450	3
RD.2.2	190	675	4½
RD.2.3	199	900	6
RD.2.4	208	1350	9
RD.2.5	215	2250	15
RD.2.6	221	5400	36

NOTE: This table consists of information selected from BS 4729 and should be read in conjunction with figure 5.71 of this section.

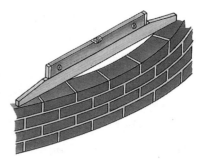

Figure 5.72. *1.2 m straight edge reaches between plumbing points.*

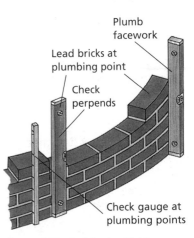

Figure 5.73. *Bedding lead bricks.*

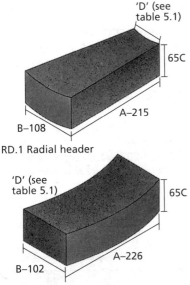

Figure 5.71. *Standard radial bricks to BS 4729.*

A BASIC METHOD OF CONSTRUCTION

First course
Carefully bed the first course of bricks to the line of the radius marked in the screed. Use the trammel and spirit plumb rule to check the alignment. Any 'kinks' in the first course will be continued to the full height of the wall.

Level-in bricks using a straight edge and spirit level *(fig 5.72)*.

Plumbing points and templates
As there are no quoins either end of a curved wall, establish plumbing points at about 1200 mm intervals. Make templates any convenient length, 1200 mm is a comfortable size to support in one hand while holding a trowel in the other. Longer templates require fewer plumbing points.

Like quoins in straight walls, plumbing points control plumb, gauge and alignment for the full wall height. Levelling of courses also takes place between plumbing points because that is where the gauge is controlled *(figs 5.72 & 5.73)*.

Straight bricks and templates
The whole length of the faces of straight bricks cannot follow a curved template. Instead, in a **convex** face, both arrises of all bricks must touch the template *(fig 5.74)*.

For a **concave** face where a reverse template is used *(fig 5.75)*, the centre of the face of each brick must touch the template, and the arrises should be equidistant from the template.

Radial bricks and templates
With radial bricks manufactured to suit the particular curvature of a wall, the whole face of each brick should touch the template.

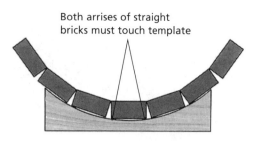

Both arrises of straight bricks must touch template

Figure 5.74. *Checking alignment of convex face with template.*

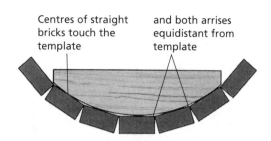

Centres of straight bricks touch the template

and both arrises equidistant from template

Figure 5.75. *Checking alignment of concave face with reverse template.*

ANOTHER METHOD OF CONSTRUCTION

Where space allows, the curvature of a wall can be controlled using a trammel only. A template is not required.

A steel rod is fixed vertically (fig 5.76) instead of a wooden peg (fig 5.68).

A trammel is used to check the position and alignment of every brick.

Temporarily concrete-in the steel rod at least one day before it is to be used and take care it is truly plumb and rigidly fixed.

At the completion of each course raise the trammel up the steel rod by the height of one course. Support the trammel with an elastic band wound round the steel rod and rolled up for each course.

This simple device supports the trammel horizontally, and allows the radius dimension to the face side of each course to be checked as it is laid.

Levelling between plumbing points must still be carried out as in *fig 5.72*.

Verticality and **gauge** must still be checked at plumbing points as in *fig 5.73* e.g. at 1 m intervals round the circumference.

THE NEED FOR CARE AND ATTENTION

To form a 'sweet curve' requires great care and attention to make courses truly level and to keep plumbing points accurate. Check that **both** ends of **every** brick touch the straight edge when levelling.

Curved brickwork will be as good as well-laid straight walling only if very great care is taken with plumbing points and templates.

References

(1) BS 4729:1990 *'Dimensions of bricks of special shapes and sizes'.*

(2) Hammett and Morton, 'The design of curved brickwork'. *The Brick Development Association.*

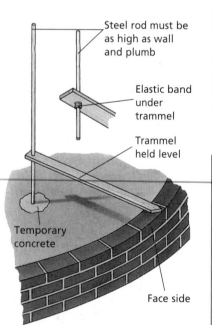

Steel rod must be as high as wall and plumb

Elastic band under trammel

Trammel held level

Temporary concrete

Face side

Figure 5.76. *Controlling curvature of each course with a trammel from a centre rod.*

KEY POINTS

- Locate exact position of striking point.
- Accurately shape templates to the correct radius.
- Set out carefully, keeping trammel horizontal.
- Lay out bricks 'dry' around the curve to check the size of cross joints.
- Make distances between plumbing points equal to length of template.

- Plumb down from datums to begin gauging at each plumbing point. *(see Section 2.2)*
- Continually check plumb and gauge, only at plumbing points.
- Pencil plumb perpends at every plumbing point. *(see Section 2.4)*
- Level bricks between plumbing points with great care.

5.8   CORBELLING

A corbel, in general building terms, is an isolated or continuous feature built into and projecting from the face of a wall to support walling, in front of the main wall, or roof trusses, floors or beams.

The unusual stone corbel in an arcade at Queen's College Cambridge *(fig 5.77)* and the accompanying diagram *(fig 5.78)* illustrate the way a corbel works in principle. The tendency for the load 'A', supported by the projecting portion, to rotate the corbel anticlockwise is counterbalanced by load 'C', from the brickwork above, acting clockwise to prevent tension and cracking developing at 'B' and possible failure of the corbel to carry its load.

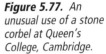

Figure 5.77. *An unusual use of a stone corbel at Queen's College, Cambridge.*

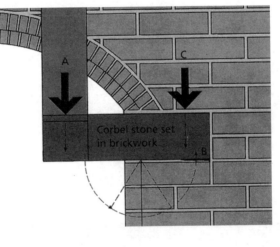

Corbel stone set in brickwork

Figure 5.78. *Demonstrating the principle of the corbel.*

TRADITIONAL BRICK CORBELLING

Traditionally, the main purpose of corbelling was to provide structural support but the opportunity was often taken to enrich the structure with mouldings or carvings. Brick corbelling may consist either of single point supports, such as a drop corbel, or of continuous oversailing courses each projecting beyond the one below.

The following two examples are included solely to describe the basic bricklaying skills needed to build traditional corbelling and from which the skills to build modern examples can be developed.

Simple structural brick corbelling must not exceed the limits recommended in the Masonry Code of Practice[1] *(fig 5.79)*, unless designed by a structural engineer.

PREPARING TO BUILD A DROP CORBEL

The corbel in this example is to support a two-brick wide attached pier, projecting from the main wall by the length of one brick *(fig 5.80)*.

The position of the pier, the bonding and perpends should have been planned at ground level before bricklaying began.

- The corbelling begins three courses below the full width and projection of the attached pier. Each course projects a quarter of a brick, both parallel with and at a

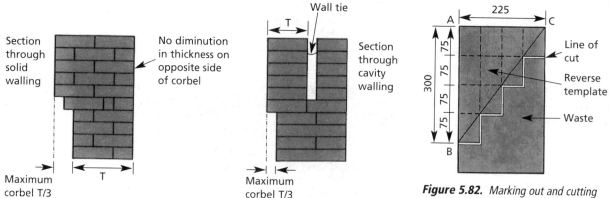

Section through solid walling

No diminution in thickness on opposite side of corbel

Maximum corbel T/3

Figure 5.79. *Recommended extent of brick corbelling.*

Wall tie

Section through cavity walling

Maximum corbel T/3

225

A ⟶ C

300

75
75
75
75

B

Line of cut

Reverse template

Waste

Figure 5.82. *Marking out and cutting reverse template.*

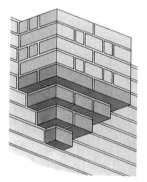

Figure 5.80. *An example of a drop corbel.*

right angle to the face of the wall.
- The first header is on the centre line projecting by a quarter-brick *(fig 5.81a)*.
- The second course of two headers projects all round by a further quarter-brick or a total of a half-brick *(fig 5.81b)*.
- The third course of three headers projects a total of three-quarters of a brick *(fig 5.81c)*.
- The final course is two stretchers wide to begin the English bond in the pier *(fig 5.81d)*.

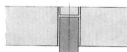

a. First course.

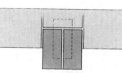

b. Second course.

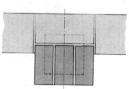

c. Third course.

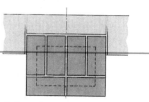

d. Fourth course.

Figure 5.81. *Plans for corbel courses.*

Keeping projections even and plumb

The projecting corbels must be equal, level, straight and parallel. A number of techniques for doing this are described below.

Making a corbelling template

- Take a piece of hardboard exactly 225 mm wide with end AC cut square to the sides. Mark off 300 mm (AB) down one side. Join BC *(fig 5.82)*.
- Divide AB into the four corbel courses including bed joints, i.e. 75 mm.
- Project these square from AB to intersect line BC. From these intersections draw lines square to AC and projecting below line BC.
- Cut along the stepped line below and to the right of BC. You now have an accurate profile of the proposed brick corbels. Use this **reverse** template to transfer the marks to a sheet of plywood and then cut the final template *(fig 5.83)*.

The template is used with a plumb level to check the profile of the corbelling *(fig 5.84)*.

KEEPING CORBELS SQUARE TO THE WALL, ALIGNED AND LEVEL

- Use a square held against the wall to check each course *(fig 5.85)*.

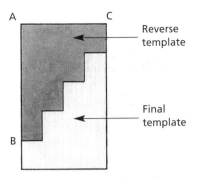

Figure 5.83. *Marking outline of final template.*

- Use a straight edge to align corbelling at each end of the pier face as well as the side returns *(fig 5.86)*. Line through the face as each course is laid.
- Level and line-in the **soffit** of each corbel course *(fig 5.87)*.

- As a double check place a straight edge along the face of each corbel course and measure its distance from the wall at each end. They should be the same *(fig 5.88)*.

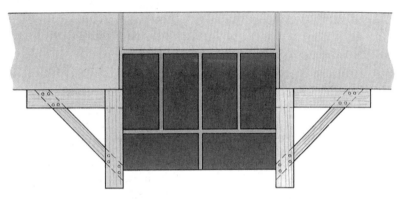

Figure 5.85. *Keeping corbels square with wall.*

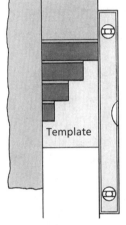

Figure 5.84. *Using a template with a plumb level.*

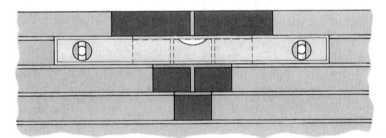

Figure 5.87. *Levelling corbel courses.*

Figure 5.86. *Aligning corbels with a straight edge.*

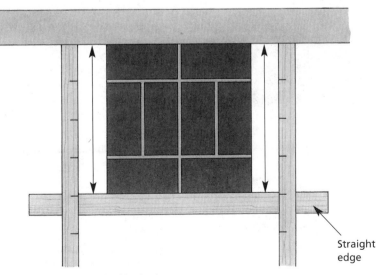

Straight edge

Figure 5.88. *Making a double check.*

BUILDING THE DROP CORBELLED COURSES

- Raise the main wall to the beginning of the corbel or corbels. Mark the position of the corbel with a header placed dry allowing a joint at each side *(fig 5.89a)*.

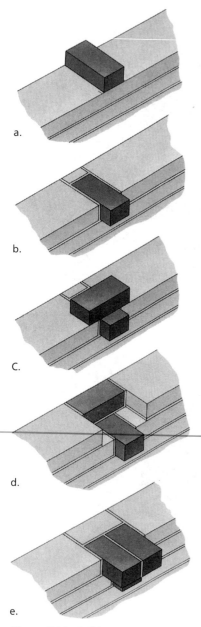

a.

b.

c.

d.

e.

Figure 5.89. *Building the corbel.*

- Rack back from the corbel both sides. Bed the first corbel brick and bed a cut brick behind it *(fig 5.89b)*.
- Mark the position of the second corbel course with a stretcher placed dry *(fig 5.89c)*.
- Rack back and bed a stretcher on the cut brick trapping the corbel brick below. This is to provide back-weight to 'tail down' the corbelled bricks and provide a counter balance to prevent overturning *(fig 5.89d)*.
- Bed second course of two headers *(fig 5.89e)*, levelling underside as *figure 5.87* and check for square as *figure 5.85*.
- Build the third and fourth corbel courses similarly, bedding cut bricks behind and solidly 'tailing down' each course of corbelling.
- When bedding corbel bricks, allow them to tilt a little 'backwards' or 'inwards' to the main wall, for as each successive course is tapped into place the bricks in the course below will settle slightly until level *(fig 5.90a & b)*.

The two stretchers in the fourth corbel course may well require temporary support by a timber batten until firmly 'tailed down' by subsequent courses.

Align and level the soffits, not the top of each brick course as is normal. The eye is always drawn to projecting features and will readily notice any unevennesss.

If perforated bricks are used in the main wall, check, before beginning work, whether purpose-made solid versions

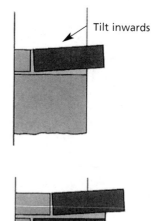

Tilt inwards

a.

Brick settled in level position as courses above are levelled

b.

Figure 5.90. *Tilting corbel bricks inward.*

have been ordered so that perforations will not be visible from below. Also, if the type of brick used has bed surfaces markedly different from the faced surfaces, a Standard Special BD 1.3[2] which has one bed surface faced may have been ordered, or even a purpose-made special that has both stretcher surfaces faced.

Frogged bricks should of course be laid frogs uppermost.

WALLING LENGTHS OF CORBELLING

Total projection from the wall face, projection of each course, setting-out, use of template and checking for squareness and alignment are carried out as described for the drop corbel.

An additional requirement is the necessary use of line and pins for running in each course of corbelled brickwork. The line must be secured to the bottom arris of the corner brick at each end of the course. As this can be fiddly, it may be worth using the small metal or plastic corner blocks or shoes which are available *(fig 5.91).*

BUILDING MODERN CORBELLING

In many modern buildings, steel or concrete frames support and restrain the external cladding of bricks and blocks. The outer leaves of brickwork often include a corbel-like shape purely as an embellishment. It does not support any load, and indeed is itself invariably supported and tied back to the structural frame by a system of steel angles, ties and brackets. This technique is described more fully in section 4.7.

Here, bricklayers cannot use the technique of 'tailing down' and often raise only a few courses and wait for the mortar to set and the ties to be effective before raising the work further.

References

(1) BS 5628:Part 3:1985 Code of practice for use of masonry.
(2) BS 4729:1990 Specification for dimensions of bricks of special shapes and sizes.

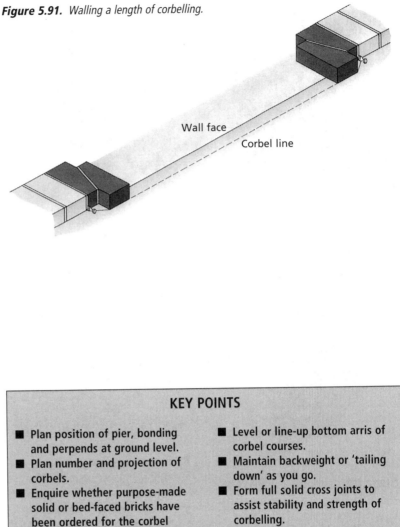

Figure 5.91. *Walling a length of corbelling.*

Wall face

Corbel line

KEY POINTS

- Plan position of pier, bonding and perpends at ground level.
- Plan number and projection of corbels.
- Enquire whether purpose-made solid or bed-faced bricks have been ordered for the corbel bricks.
- Keep all projections equal.
- Keep all corbel courses parallel.
- Level or line-up bottom arris of corbel courses.
- Maintain backweight or 'tailing down' as you go.
- Form full solid cross joints to assist stability and strength of corbelling.
- Carefully joint up underside of corbel courses.

5.9 TUMBLING-IN COURSES

Tumbling-in is a method of reducing the plan section area of brickwork between two levels by providing a sloping weathered surface in place of the horizontal surface or ledge which would otherwise result. Tumbling-in was commonly used in the past to reduce the width of an external chimney breast to that of the chimney and to reduce the projection of attached piers *(figs 5.92 & 5.93).*

The skills required should not be lost by neglect merely because the feature has been little used in recent years. Architects might more often find tumbling-in appropriate if they were aware that bricklayers still have the necessary knowledge and skill.

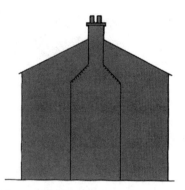

Figure 5.92. *Tumbling-in to reduce an external chimney breast to a chimney.*

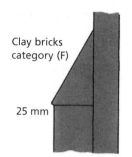

Figure 5.94. *First course is commonly projected.*

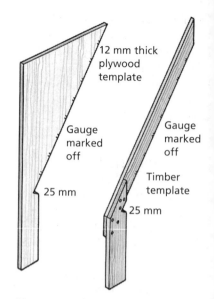

Figure 5.95. *'Gun' templates to maintain slope of short lengths of tumbling-in.*

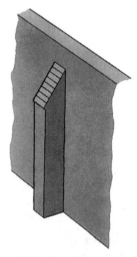

Figure 5.93. *Tumbling-in to reduce an attached pier.*

should be specified as for walls with a capping. (*See Section 6.2 'Frost attack and frost resistance', table 6.5 of Section 6.4 'Durability of brickwork' for recommended brick and mortar designations and Section 5.1 'Copings and cappings' for definitions and comments on cappings and copings*)

TWO METHODS OF BUILDING

1. For small reductions
Tumbling-in is set-out full size on a board to determine the size and shapes of cut bricks. A timber template is made and used to maintain the required slope *(fig 5.95)*.

2. For large reductions
Tumbling-in courses are built using lines set to the required angle. Regularity of gauge is very important (i.e. 4 courses to 300 mm) and is maintained by using a standard gauge rod to check each course of tumbling-in as these are set in position.

In both methods a bricklayer's level should be used in conjunction with a try square to obtain true, horizontal bedding cuts to the bricks *(fig 5.96)*.

CHOICE OF BRICKS
The faces of the bricks form the sloping weathered surface and it is usual to project the first course of tumbling because it is impractical to cut bricks to a sharp edge along the exposed angle and difficult to finish the mortar joint satisfactorily *(fig 5.94)*.

The bricks and mortar in the weathered surface should be specified as for a capping. Because the projection is not throated, the bricks and mortar below the weathered surface

Figure 5.96. *Bevel is set once to cut all bricks used in tumbling-in.*

Clay engineering and perforated bricks are best cut with a bench-mounted masonry saw for neatness, accuracy and reduced wastage. Frogged bricks should be cut to avoid a 'vulnerable' end by forming the long side on the solid bed rather than on the frogged bed. *(see Section 2.5 'Cutting bricks')*

ILLUSTRATED EXAMPLES
1. For small reductions
Half-brick deep bonding pocket formed to receive tumbling-in courses *(fig 5.97)*.

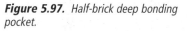

Figure 5.97. *Half-brick deep bonding pocket.*

A gauge rod is used to compare a **whole** number of courses of tumbling-in with a **whole** number of vertical courses. In this case 8 and 7 courses respectively give a slope of approximately 60 degrees on a pier which projects 1½ bricks (337 mm) *(figs 5.98 and 5.99)*.

Every time the 'gun' template is used press the stem firmly against the plumb face of the attached pier checking
(a) Slope of face.
(b) That tumbling-in courses are at right angles to template, using a try square.

2. For large reductions
Bricklaying line fixed each side of a longer length of tumbling-in where a template would be inappropriate *(fig 5.100)*.

Gauge rod

Spacer block

Figure 5.98. *Use of 'gun' template.*

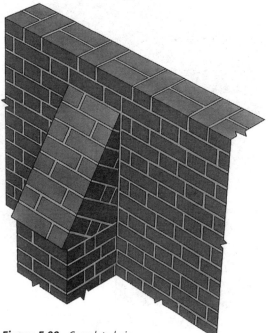

Figure 5.99. *Completed pier.*

Optional thin plywood temporarily fixed each side to keep tumbling-in vertical

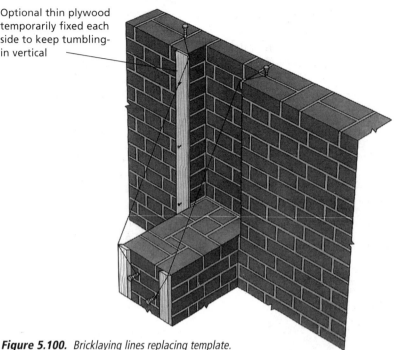

Figure 5.100. *Bricklaying lines replacing template.*

Other examples

- Methods of using tumbling-in courses to finish a sloping freestanding wall or the balustrade wall to an external staircase are shown in *figs 5.101 & 5.102.*
- The solutions shown in *figs 5.101 & 5.102* might be adapted to finish a gable parapet wall.
- Method of tying in a brick-on-edge capping to check the tendency to slide downwards on the bed *(fig 5.102).*
- Reducing a chimney breast on an elevation to the size of the external chimney stack *(fig 5.103).*

Figure 5.101. *Tumbling-in to a sloping freestanding wall.*

Figure 5.102. *Tying in a brick-on-edge capping.*

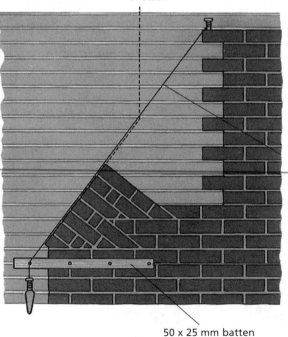

Ultimate position of stack

Line and pins to maintain correct angle

50 x 25 mm batten

Figure 5.103. *Reducing a chimney breast.*

5.10   FIREPLACE OPENINGS, CHIMNEY BREASTS AND FLUES

In modern houses heating is provided by central heating installations and consequently many do not have a fireplace or a chimney. However, an open fire is widely regarded as an attractive feature and fireplaces are now often being built in modern houses as luxury features. They are often embellished with decorative detail and so present a rewarding opportunity to demonstrate a bricklayer's skill and competence *(fig. 5.104 and 5.105)*.

Regardless of the detail, design and visual attraction of a fireplace it will need to function effectively. To ensure this the bricklayer must appreciate the technical requirements of fireplace, chimney and flue installations and the Building Regulations that apply to them. The construction should be completed in accordance with the requirements and Regulations as well as demonstrating neatness and skill.

The design and construction of fireplace openings, chimney breasts and flues are controlled by Part L of the Building Regulations. Further advice and guidance can be found in BS 6461: 'Installation of chimneys and flues for domestic appliances burning solid fuel' and in Part 1 of these regulations: 'Code of practice for masonry chimneys and flue pipes'.

Figure 5.104. *Decorative freestanding, double-sided fireplace.*

Figure 5.105. *Modern fireplace with firewood compartment alongside.*

These notes relate to the installation in domestic buildings of fireplaces, flues and chimneys for heating appliances up to 45 kW output.

TYPES AND POSITION OF FIREPLACE OPENINGS

The Building Regulations that deal with fireplaces, chimneys and flues are mainly concerned with the avoidance of the spread of fire to the surrounding structure and with the safe discharge of the products of combustion to the atmosphere.

In order to comply with these Regulations open fires have to be contained within a fireplace recess. Careful consideration must be given to its position and construction.

Types of fireplace
1. **Chimney breast built to an external cavity wall**
   (a) Chimney breast positioned within the room *(fig 5.106)*.
   (b) Chimney breast built externally (also known as a 'reardorse') *(fig 5.107)*.
   This type has the advantage of not reducing the floor area of the room.
2. **Fireplace openings placed on internal walls**
   (a) Single fireplace *(fig 5.108)*.
   (b) Back-to-back fireplace openings *(fig 5.109)*.
   This arrangement is quite often used in semi-detached and terraced housing so that flues can be collected together to form one chimney stack.

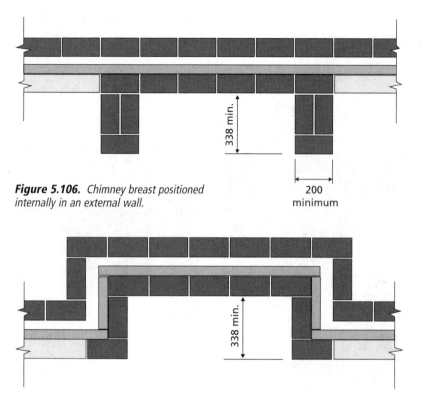

Figure 5.106. *Chimney breast positioned internally in an external wall.*

Figure 5.107. *Chimney breast positioned externally in an external wall.*

3. Other types

Other designs may involve corner sited openings, freestanding and 'through' (double-sided) fireplace arrangements (*fig 5.110*).

SIZES OF FIREPLACE OPENING

In order to accommodate a particular heating appliance, the fireplace opening should be built to specified dimensions. The width, height and depth of the opening will depend on the particular appliance or fireplace accessories to be accommodated, but to comply with Regulations the opening must be at least 338 mm (1½ bricks) deep and the jambs must to be at least 200 mm (1 brick) wide.

As the fireplace and flue are considered to form a structural

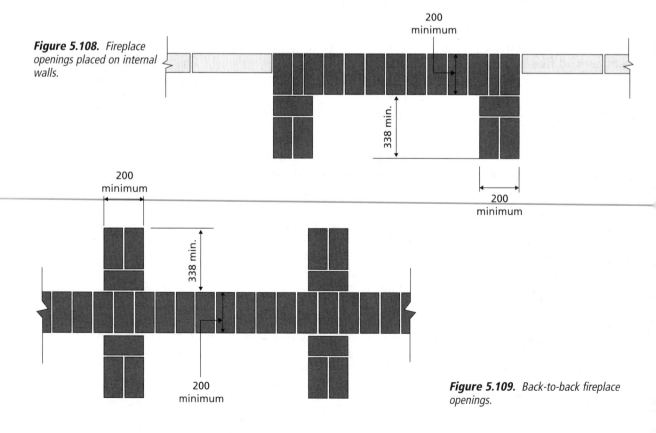

Figure 5.108. *Fireplace openings placed on internal walls.*

Figure 5.109. *Back-to-back fireplace openings.*

DEFINITIONS

The following definitions are particular to fireplaces, chimneys and flues and are in addition to those for more general terms listed in the Glossary of Terms included in this book (p. xi). They are taken from BS 6461.

Chimney   a structure (including any part of the structure of a building) enclosing or forming part of a flue or flues other than a flue pipe, including any opening therein for the accommodation of an appliance, but excluding the flue terminal

Chimney breast   a projection beyond the thickness of a wall containing the fireplace and flue(s)

Chimney connector   an accessory that connects an appliance or flue pipe to a chimney

Chimney jamb   the walling at the side of a fireplace recess

Chimney stack   part of a chimney enclosing one or more flues that rises above the roof of a building of which it forms a part and which includes the chimney terminal, but not the flue terminal

Chimney terminal   the uppermost part of a chimney stack

Fireplace recess   a space formed in a wall or chimney breast into which an appliance may be placed and from which a flue leads

Flaunching   the weathering formed in mortar at the top of a chimney or base of a flue terminal

Flue   a passage that conveys the products of combustion from an appliance to the open air

Flue block   a factory-made masonry unit that can be erected on site to form a chimney. It may contain voids for either insulation or for combustion air

Flue lining   a lining forming the wall of a flue for the purpose of protecting the chimney fabric

Flue pipe   a pipe used for connecting the appliance flue outlet to the chimney flue or to the outside atmosphere, but not including a pipe used as a lining to a chimney

Flue terminal (chimney pot)   a prefabricated or built-up unit forming the outlet end of a flue

Gather (oncome)   the contraction over a fireplace recess to reduce it to the size of the flue

Lintel   a load-bearing and/or throat-forming beam above the fireplace recess

Offset   a double bend introduced into a flue so that its direction remains parallel to its original direction. The effect is to give the path of the flue a lateral displacement

Oversailing   courses of stone or brickwork (masonry) arranged to project from the face of a wall or chimney stack largely for decorative effect

Throat   that part of the flue, if contracted, which is located between the fireplace and the chimney flue

Withe (midfeather, bridge, brig)   a partition between adjacent flues in a chimney

Figure 5.110. *Circular freestanding fireplace with raised hearth.*

part of the building all walls must be taken down to foundation level and a horizontal DPC provided to resist moisture rising up into the superstructure. The DPC will need to link with DPCs and/or DPMs adjacent to the fireplace opening.

Figures 5.111a & b illustrate a typical design and minimum dimensions.

THE CONSTRUCTIONAL HEARTH

To contain a fire safely every fireplace must have a constructional hearth at its base.

This is a concrete slab 125 mm thick (minimum) that must extend into the full depth of the opening and project in front of the face line of the chimney jambs a minimum of 500 mm. It must also extend a minimum of 150 mm on either side of the opening.

BRIDGING THE FIREPLACE OPENING

The fireplace opening must be closed over at the top and the brickwork over it must be supported. Flue liners in the chimney above also require

a.

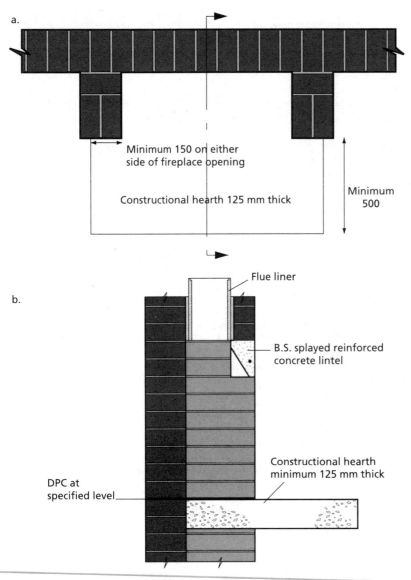

Minimum 150 on either side of fireplace opening

Constructional hearth 125 mm thick

Minimum 500

b.

Flue liner

B.S. splayed reinforced concrete lintel

Constructional hearth minimum 125 mm thick

DPC at specified level

Figure 5.111. *a: Plan and b: Section of typical fireplace design showing minimum dimensions required by regulations.*

Splayed corbels tied into chimney breast

Figure 5.113. *Arrangement of splayed corbels from chimney breast to 'gather' opening into flue.*

fireback – a prefabricated unit of fireproof material that lines the back and sides of the fireplace. This restriction improves the flow of flue gases from the fireplace and up into the chimney. The lintel is formed with 'wall holds' (squared ends) to assist bedding it in the jambs.

In conjunction with the splayed lintel the brickwork on each side of the opening behind it, has to be 'gathered over' to form a seating for the flue liners above; this has to be built by the bricklayer by forming a number of splayed corbels ('oncomes') on either side until the opening is reduced to a width to suit the flue liner *(fig 5.113)*.

PRE-CAST THROATING UNIT

Prefabricated pre-cast concrete throating units are available as an alternative to the splayed lintel, fireback and corbelling described above. They are manufactured to suit standard plan dimensions and have three functions:

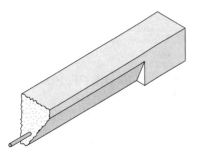

Figure 5.112. *Isometric rear view of B.S. splayed reinforced concrete lintel.*

support. A British Standard splayed reinforced concrete lintel is specifically designed and manufactured for this purpose *(fig 5.112)*.

In addition to providing support for the brickwork and flue liners over the opening, the splay at its back forms a restricted throat for the flue way when used in conjunction with a

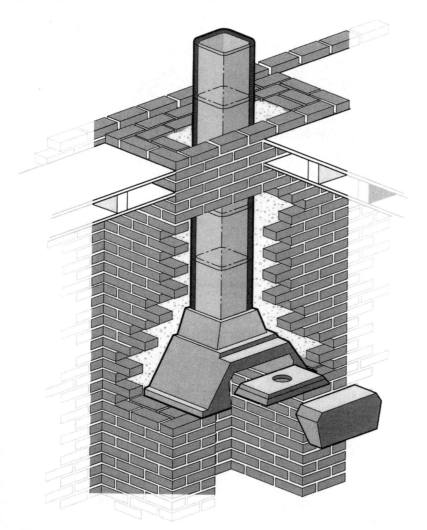

Figure 5.114. *Proprietory pre-cast throating unit.*

(i) bridging the fireplace opening and providing support for brickwork above
(ii) forming part of the restricted throat requirement
(iii) gathering the fireplace opening to provide a structural seating for the first flue liner. Alternative fittings provide for circular or square liners *(fig 5.114).*

FLUE LINERS

Before the 1960s chimneys were very rarely lined, instead they were internally rendered with lime mortar parging, or pargeting. With the advent of modern solid fuels chimneys became more vulnerable to damage by the products of combustion, as solid fuels contain greater concentrations of carbon dioxide and sulfur. As flue gasses reached the colder air at the top of the chimney stack, condensation of water vapour would occur. The condensate, running back down the flue, dissolved carbon dioxide and sulfur to form a mixture of weak carbonic and sulfuric acids. Over a number of years this acid would seep into and attack the lime mortar of the parging and the joints of the brickwork, causing it to disintegrate and fall down the flue.

To avoid this problem the Building Regulations made the installation of flue liners compulsory *(figs 5.115 & 5.116).*

Purpose-made flue liners are manufactured from clay or refractory concrete, both of which are non-combustible and are resistant to damage by flue gas condensate.

Liners are manufactured in various sizes and are either circular or square in section. They have rebated or socketed joints to provide an effective seal and it is very important that the bricklayer installs them the correct way up otherwise flue gas condensate can seep through the joints and cause damage in the brickwork beyond. Liners must always be fixed with sockets or rebates facing up to receive the plain or spigoted end of the next liner.

When starting to build a flue, first check that every flue liner is sound and fit for its purpose. They should all be free from cracks and splits, and the ends should not be damaged.

In all cases the flue liners should be positioned ahead of the brickwork being built around them. Proper joints should be made between the liners, usually in the same mortar as the rest of the chimney construction. All joints between the liners should be full and finished to leave a clean internal bore inside the flue.

Weak lime mortar or insulating concrete

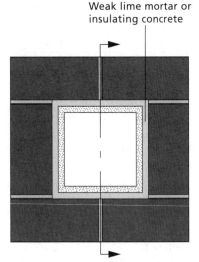

Plan view of chimney stack

Figure 5.116. *Flue liners are manufactured straight and curved.*

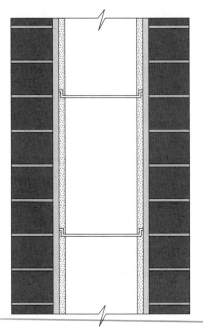

Section through lined flue

Figure 5.115. *Modern flues are lined with clay or refractory concrete flue liners (plan and section).*

The space between the outside of the flue and the surrounding brickwork should be filled with a suitable material that will allow expansion to take place, with no resultant damage. Suitable materials are:

1) A weak lime/sand mortar
2) Insulating concrete (usually a mix of exfoliated vermiculite granules with a small amount of ordinary Portland cement).

THE DESIGN AND CONSTRUCTION OF FLUES

The design of houses up to 50 or 60 years ago encouraged the use of fireplaces in every room of the building. Flues built in brickwork would travel in various directions in order that they might be collected together to form a chimney stack. In modern construction it is common for only one fireplace, if any, to be constructed within a dwelling and therefore only one flue is required.

Flue travel

In many cases nowadays, architects will specify that the vertical line of the flue should be offset by at least the dimension of its width, although there are no proven benefits for doing this.

This offsetting of the flue is termed 'flue travel' and is required when the collection of flues to form one chimney stack is necessary.

To achieve the desired travel, purpose-made liner bends are used. The flue can be moved in any direction, but the angle of the flue should be no more than 30° from the vertical *(fig 5.116)*.

The design and detail of fireplace openings, chimney breasts and flues may differ considerably according to location and the heating appliance to be accommodated. However, the bricklayer should appreciate and understand the basic principles involved in their construction as they apply to all designs and need to be properly implemented for the successful functioning of the installation.

References

- BS 1251 (1987) Specification for open-fireplace components.
- BS 6461: Installation of chimneys and flues for domestic appliances burning solid fuel.
- Part 1 (1984): Code of practice for masonry chimneys and flue pipes.
- BS 1181 (1989) Specification for clay flue liners and flue terminals.

KEY POINTS

- Check that all special fireplace components to be built in are to hand, are the required size and are undamaged.
- Check that a sufficient number of the flue liners are to hand and that they are undamaged.
- Check the proposed level for DPC with regard to continuity with adjacent DPCs and DPMs in floor construction.
- Locate and set-out the constructional-hearth accurately.
- Locate and set-out fireplace opening and chimney breasts accurately.
- Continually check line, level, plumb and gauge.
- Take particular care to fix the flue liners the correct way up (socket facing up).

- If the fireplace and chimney are to form a decorative brickwork feature check the Key Points listed in Section 5.6 'Decorative brickwork'. Also those in 5.3 'Curved arches' and 5.5 'Soldier arches' as appropriate.
- Select bricks for exposed internal face work with particular care as work of this nature is likely to get close and frequent scrutiny.
- Keep work clean and protect partially completed work during breaks.
- Protect the whole feature on completion while awaiting handover.

5.11 CHIMNEY STACKS FOR DOMESTIC FIREPLACES

The construction of chimney stacks and flues is quite a complex task, calling for the integration of work by craftspeople dealing with carpentry, roofing, plumbing and bricklaying. The bricklayer needs to grasp the overall requirements of this work and ensure that they are applied correctly.

This section complements the previous one about the construction of fireplaces and flues. It deals with the siting of a chimney stack and its construction as it passes through the roof surface and into the open air above. It also covers detail of the design and construction of the chimney terminal.

Emphasis is placed on the need for damp-proof course trays and flashings to prevent rainwater penetration where the stack penetrates the roof and it is essential that the bricklayer has a good working knowledge and understanding of the correct installation of these.

POSITION OF CHIMNEY OUTLETS

The position and height of a chimney outlet can play a major part in the efficiency of a fireplace and flue.

It is very important that the chimney terminates in a position where the products of combustion do not become either a health hazard or a fire risk. Building Regulations require that flue gases should be discharged out of the building clear of any window, rooflight or other openings and sufficiently far from any material which might ignite if it was to be in contact with hot flue gases or sparks.

Wind and the effects of adjacent structures, trees, etc., cause zones of high and low pressure about buildings. A low pressure zone would generally occur on the lee side of the ridge

a.

Figure 5.117. *Historically chimneys have often been major decorative brickwork features especially in Tudor architecture. a: Hampton Court Palace, b: Rye House, Essex.*

b.

Figure 5.118. *Chimneys are also a feature in recent domestic architecture.*

of a pitched roof, and close to the windward side of a flat roof (10° or less is considered to be flat). Corresponding zones of high pressure are developed on the opposite sides of the roof. The position and design of a chimney outlet should be considered in relation to these pressure zones as they may seriously affect the efficiency of the flue.

Theoretically a flue is likely to function well if the outlet is in the low pressure zone as the flue gases would be drawn out into the atmosphere. Flue gases would be driven back into the building if the outlet was in a high pressure zone. Because the direction of the wind varies at different times, pressures about a roof surface change and therefore chimney outlets should be sited sufficiently above a roof surface to avoid these pressure effects.

To satisfy normal requirements for the effective and safe working of flues their outlets **should not** be sited within the shaded areas shown in *figs 5.119a & b.*

HEIGHT AND STABILITY OF CHIMNEY STACKS

To ensure the stability of chimney stacks, the Building Regulations require that the height of an unrestrained chimney should not be more than 4.5 times its least width at the level it penetrates the roof surface *(fig 5.120).*

This limitation is likely to restrict the use of a chimney of minimal dimensions (440 mm – 2 bricks) in steeply pitched roofs unless they are located at or near the roof ridge. For example, the diagrams indicating the required heights of flue outlets show that a chimney terminal must be above any part of a roof surface within a horizontal distance of 2.3 m; therefore a brickwork chimney located at

the eaves of a 45° pitched roof would have to be a minimum of 2.3 m high and therefore its minimum width would be 565 mm (2½ bricks).

For tall stacks special structural design might be necessary, perhaps incorporating horizontal bracing to improve stability.

CONSTRUCTION OF CHIMNEY STACKS

All chimney stacks, whether as a single flue or a number of flues grouped together, should be constructed in such a way that every flue is surrounded by at least 100 mm of solid masonry. Each flue must serve only one fireplace or heating appliance.

A chimney stack is one of the most exposed parts of any building and therefore appropriately durable materials should be selected and the highest standard of workmanship maintained throughout its construction.

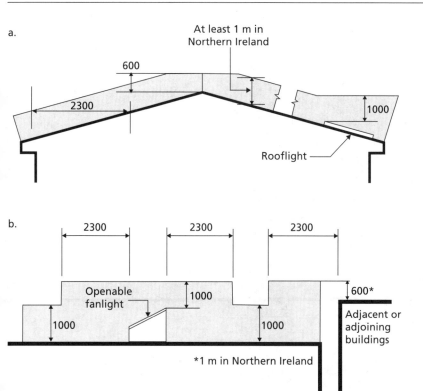

Figure 5.119. *Flue outlet should terminate outside of the shaded areas. (a) Roof with a pitch of 10° or greater. (b) Roof with a pitch of less than 10°.*

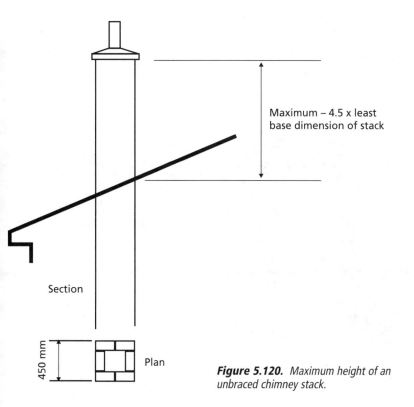

Figure 5.120. *Maximum height of an unbraced chimney stack.*

THE TOP OF A CHIMNEY STACK

Due to its exposed position at the top of the stack the flue terminal, or chimney pot, should be bedded into at least 3 to 4 courses of brickwork. This will ensure that it is well anchored and not likely to be dislodged by strong winds.

The top of the stack may incorporate oversailing courses which are frequently more decorative in nature than functional. The top surface of the chimney surrounding the flue terminal, or chimney pot, may consist of a sand and cement weathering, or flaunching, to drain away rainwater rapidly in wet weather. Alternatively, and providing more effective protection than flaunching, an impervious, weathered, overhanging and throated coping may be installed. If the coping is jointed, it should be set on a DPC sandwiched in mortar. This will shed rainwater quickly, throwing it clear of the stack below.

WATER PENETRATION AND CHIMNEY STACKS

The detail design and construction of a chimney stack should be considered in relation to the prevention of penetration by rainwater:

1) passing through the joint between the roof surface and the masonry of the chimney stack
2) entering the masonry above roof level and percolating down within it, bypassing the flashings at the roof/chimney junction

Figure 5.121. A variety of clayware flue terminals.

Figure 5.122. Chimney terminated with dentils, oversailing courses and flaunching around chimney pots.

Figure 5.123. Chimney terminated with pre-cast coping, weathered and throated to give good weather protection.

3) penetrating the masonry of the stack to the surface of the flue liner and from there running down to lower levels within the building.

Lead DPC trays and flashings are required and these are usually formed by a plumber. Trays and DPCs are installed by the bricklayer but generally the plumber fits the flashings. However, the bricklayer needs to have a good working knowledge of the total assembly and how an effective construction is built. Good design points are illustrated in *fig 5.124* and include:

1) Chimney abutment flashings comprising stepped and lapped raking flashings, a back gutter and flashing and an apron flashing. These provide a barrier to water entering the joint between the roof surface and the chimney stack.

2) A pre-formed DPC tray with upstands at the back and two side edges and a projection at the front to dress down over the abutment flashing to be fitted below. The tray should be taken through a joint between the flue liners and dressed up on the inside of the liners a minimum distance of 25 mm. This tray is to prevent water that may penetrate the masonry above the roof from moving down within it and into the building. It is built into the stack where the chimney intersects with the roof plane at its lowest point.

3) In steeply pitched roofs, and particularly where severe exposure to wind-driven rain

Figure 5.124. *Section through chimney showing good design features.*

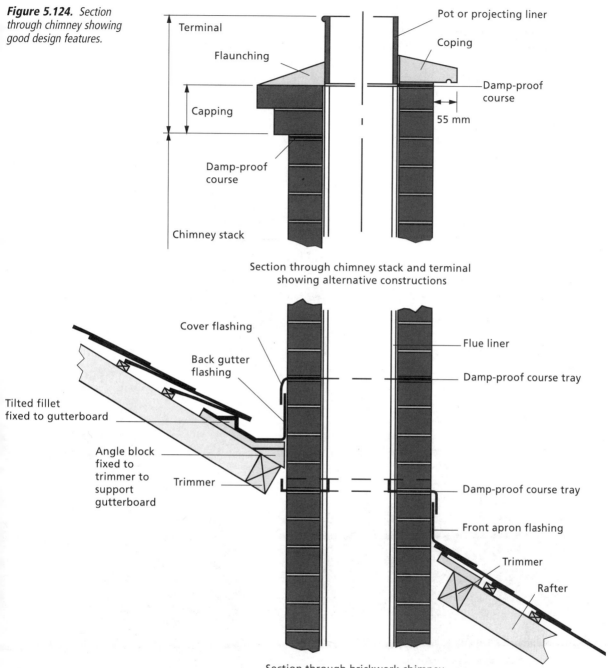

Section through chimney stack and terminal
showing alternative constructions

Section through brickwork chimney

is anticipated, an additional, but simpler, DPC tray is installed at the higher level chimney/roof plane intersection. This is intended to reduce the amount of water that may reach the

lower tray. Opinions vary about the correct positioning of DPCs and trays in chimney stacks and some architects and/or local authorities may prefer a variation of the above recommendations.

To prevent a risk of its corrosion when in contact with mortar, lead built into a chimney stack as a tray should be coated with a solvent-based bituminous paint on both sides before installation. Flashings do not need a bituminous coating as they are not built into the mortar joints of the brickwork by more than 25 mm and therefore are not vulnerable to corrosion.

CONSTRUCTION

Stage one: Raise the brickwork up to two courses above the lower point of intersection with finished roof and bed the specially-shaped DPC tray with upstand edges *(fig 5.125a)*.

To enable the tray to continue into the flue, and be dressed up on the inside of the liners a minimum distance of 25 mm, a joint in the flue liner must coincide with the position of the DPC tray. This joint should be a butt joint (not a socket) and may require a specially cut liner to suit.

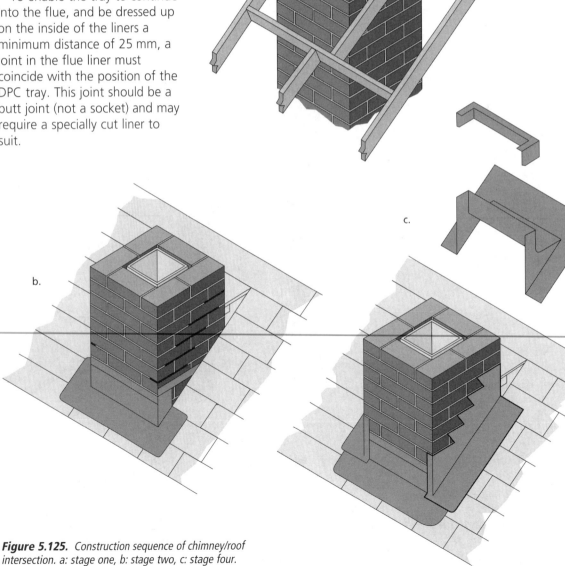

Figure 5.125. *Construction sequence of chimney/roof intersection. a: stage one, b: stage two, c: stage four.*

Stage two: Continue building the stack, leaving the mortar joints suitably raked out in readiness to receive lead flashings *(fig 5.125b)*.

If there is uncertainty about the exact position of the joints to be raked out, it is better to rake out too many joints than to have to cut out hardened mortar later; they can be pointed if not required. If a plumber is available seek advice.

Stage three: Bring brickwork up to two courses above the back gutter level and lay the second DPC tray.

Stage four: Continue building the stack to the required height *(fig 5.125c)*.

Stage five: Form any oversailing courses and complete the terminal as specified.

References
(1) BS 6461: Part 1 (1984) Code of practice for masonry chimneys and flue pipes.
(2) The Lead Sheet Manual (1990). The Lead Sheet Association.

Figure 5.126. *Chimney completed awaiting flashings to roof.*

Figure 5.127. *Not all chimneys are square.*

KEY POINTS

■ Check types of bricks and mortar are as specified or recommended.

■ Check mortar mixes are as specified or recommended.

■ Check that proposed height and location of chimney terminal conform to regulations with reference to distance above roof surface and any roof openings.

■ Check that the minimum width of the stack is adequate for the proposed height.

■ Check that the pre-formed lead trays have been correctly made to size and configuration and that there is a protective bituminous coating on both sides.

■ Check that flue liners are laid the correct way up, i.e. with sockets or rebates pointing up (see Section 5.10 'Fireplace openings, chimney breasts and flues').

■ Check that a flue liner joint coincides with the position of the DPC tray. A specially cut liner to suit may be required.

■ Bed DPC trays and the DPC at the terminal on fresh mortar.

■ Leave mortar joints recessed to suit roof abutment flashings.

■ Check that coping and chimney pot, or flue terminal, are available for installation.

6 BACKGROUND TOPICS

This section contains articles relating to technical aspects of brickwork performance. They are intended to help the reader understand the reasons why some of the details of specification and construction exist and why it is important to comply with good practice guidance in the assembly of the work.

Also included in this section are articles on understanding drawings, the care and use of tools and the manufacture of bricks.

6.1 EFFLORESCENCE AND LIME STAINING

An understanding of the causes of efflorescence and lime staining enables bricklayers, brickwork supervisors and designers to play an appropriate part in minimising the risk of their occurrence.

Figure 6.1. Efflorescence.

Figure 6.2. Lime staining.

DEFINITIONS

Common efflorescence is a deposit of soluble salts left on the surface of brickwork when the water in which they were dissolved evaporates *(fig 6.1).*[1]

Lime staining is a deposit of lime left on the face of brickwork when the water in which it was dissolved evaporates *(fig 6.2).*

Because these two undesirable occurrences are superficially similar but have different causes and manifestations and require different actions to prevent and treat them, they are dealt with separately.

EFFLORESCENCE

The most common form of efflorescence is an unsightly but harmless white deposit which does not affect the durability of the brickwork and normally disappears from new buildings within the first few months. The very rare forms of efflorescence which may cause physical harm are beyond the scope of this section.

THE SOURCES OF SOLUBLE SALTS

Common efflorescence derives mainly from soluble salts that are contained in clay bricks and sands used for mortars but sometimes in other sources.

Clay

Salts in clay are most commonly, sulfates of sodium *(Glauber's salt)*, potassium *(sulfate of potash)*, magnesium *(Epsom salts)* and calcium *(gypsum)*. Ferrous sulfate is not common in clays used for brickmaking today but where it occurs it may be responsible for rusty stains on mortar or bricks.

Mortar sands

Most sands come from pits or river beds and contain few salts. Sea sands contain many harmful salts and should not be used for mortars unless they have been effectively washed by a reliable supplier.

Cement

Fortunately, in Britain, Portland cement is most commonly used and its contribution to efflorescence is minimal. Some of the slag cements used abroad contain appreciable quantities of sodium sulfate which is a common cause of efflorescence and this should be considered before using such cements.

Detergents used as plasticisers

Many detergents contain sodium sulfates and should under no circumstances be used in place of properly formulated proprietary mortar plasticisers. *(see Section 4.1 'Mortars')*

Other sources

Bricks may absorb salts from ashes or the soil on which they are standing or from materials stacked or heaped in contact with them.

How efflorescence forms

Water dissolves soluble salts in

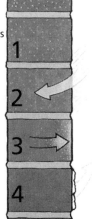

Brickwork contains soluble salts

Brickwork saturated – dissolves salts

Brickwork dries out – solution migrates to surface

Water evaporates – soluble salts crystallise on surface

Figure 6.3. *Efflorescence – a simplified diagram.*

the bricks and mortar and holds them in solution. As the water evaporates and the wall dries out the solution becomes more concentrated until salts begin to be deposited. This may occur out of sight within the pores of the brick or on the surface.

Exactly where, when and how much efflorescence will occur, is difficult to predict being dependent on complex chemical and physical conditions such as the type of salts, the rate of drying by wind and sun and the degree of saturation of the brickwork. The latter is the only condition which we can readily control.

Minimising the risk of efflorescence

The risk cannot be entirely avoided as it is not practicable to eliminate all salts from bricks and mortars nor keep them virtually dry, but it can be minimised by reducing the amount of water penetrating the brickwork.

Design details

Designers can use 'umbrella' details which protect the

brickwork from saturation. Such details include effective roof verges and eaves, copings and sills to shed run-off water clear of the brickwork below *(see Section 5.1 'Copings and cappings')*. They should also avoid details which shed water on projecting plinths and other features.

Site practice

Bricklayers and site supervisors are responsible for good site practice. Bricks should be stacked clear of the ground on pallets or by other means. Mortar materials should be kept free from contamination, and bricks in stacks, on the scaffolding and newly built brickwork protected from saturation, particularly during the seven days after bricklaying. *(see Sections 1.2 'Protection of newly built brickwork'; 4.1 'Mortars')*

Scaffold boards adjacent to the brickwork should be turned back during rain to avoid splashing causing a band of efflorescence as well as mortar stains.

Damp conducted across mortar encrusted wall ties has been observed to cause efflorescence on internal facework.

Choice of bricks

Although the current BS 3921[2] requires manufacturers to classify clay bricks according to soluble salts content by a standard laboratory test on samples of individual bricks, it is important to realise that the results of the test have little if any correlation with the liability of brickwork to effloresce in practice.

Efflorescence can be avoided, or at least minimised, by appropriate design detail and good site practice as described above.

Treatment of efflorescence

It is best to allow efflorescence to weather away naturally. Do not use acid treatment, as apart from the danger to people and materials by inexpert use some salts such as vanadium may be fixed permanently. Light brushing with a bristle, never a wire, brush may be allowed.

Internal efflorescence may be treated by dissolving small patches with very little water, say a fine mist spray. The surface may then be dried with an absorbent material but, as some users have reported that this sometimes causes a blotchy appearance, it is advisable to experiment on a small inconspicuous surface at first. The absorbent material must be constantly renewed or washed free of salts.

LIME STAINING

Lime which stains the face of brickwork is often derived from the Portland cement in mortar or adjacent concrete as well. Although it might seem an obvious source, the hydrated lime in some mortars does not seem unduly to increase the liability of the brickwork to lime staining.

How lime staining forms

When newly built brickwork becomes saturated in its early life and as the Portland cement sets (*hydrates*) it releases lime into solution. On drying out, calcium hydroxide is precipitated at the

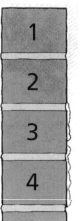

Unprotected newly built 'green' brickwork becomes saturated

Portland cement sets and releases lime into solution

Brickwork dries out – calcium hydroxide precipitated on surface

Calcium hydroxide converts to insoluble calcium carbonate

Figure 6.4. *Lime staining – a simplified diagram.*

surface (especially on mortar joints) and converts to insoluble calcium carbonate by reaction with carbon dioxide from the atmosphere. Lime staining may, of course, be accompanied by efflorescence salts.

Minimising the risk of lime staining

It is absolutely essential to protect newly built 'green' brickwork from saturation during the first 24 hours and very

important to maintain that protection for at least 7 days. (*see Section 1.2 'Protection of newly built brickwork'*)

Brickwork should be separated by a damp-proof membrane from contact with concrete that may become wet, in order to prevent the transfer of lime and salts in solution from the Portland cement in the concrete through the brickwork to the face or by washing directly over it. As there is much more Portland cement in concrete than in mortar the staining from concrete can be particularly heavy.

Treatment of lime staining

Unlike common efflorescence, lime staining is insoluble and can be removed only by expert and expensive treatment.

References

(1) Definition of efflorescence derived from 'Brickwork: Efflorescence. A perennial problem re-examined' Structural Clay Products Ltd. 1974, (out of print) by B. Butterworth, B.Sc., ARICS, F.I.Ceram., formerly on the scientific staff of the Building Research Station.
(2) BS 3921:1985 'Clay Bricks'.

KEY POINTS

- Do not stack bricks on soil or ashes.
- Protect stocks of bricks, sands and mortars from contamination and saturation.
- Do not use sea sands for mortar unless effectively washed.
- Slag cements may contribute appreciably to efflorescence.
- Use only proprietary mortar plasticisers, never detergents.
- Protect 'green' brickwork from saturation to prevent lime staining.
- Continue to protect the top of incomplete brickwork to minimise efflorescence.
- Separate brickwork from potentially wet concrete with a DPM.
- Efflorescence may derive from soluble salts in the brick, mortar materials or outside sources.

6.2   FROST ATTACK AND FROST RESISTANCE

Brickwork is at risk from frost damage as the temperature falls below freezing only if it is saturated at the time.

The risk can be minimised by designing and building to avoid saturation or by using frost resistant bricks and mortar.

FROST ATTACK – THE CAUSES

- When materials are saturated their pore structures are virtually filled with water.
- When water freezes it attempts to expand. If enclosing materials are unable to resist the stresses they will be disrupted, an action known as frost attack leading to frost damage.
- Materials liable to frost damage will be at greater risk the closer they become to being saturated and the more often the temperature falls below freezing while they are saturated. Freezing temperatures alone do not result in frost attack.
- Bricks which suffer frost attack may crumble or the face may spall away.
- Mortars which suffer frost attack will lose strength, adhesion to the bricks and be liable to erosion.

FROST RESISTANCE OF BRICKS

The resistance of bricks and mortars to frost attack, the importance of workmanship and the way design can limit saturation are described below.

Fired-clay bricks

Manufacturers are required by BS 3921[1] to classify their bricks into one of the following categories.

Frost resistant (F)

Durable even when saturated and subjected to repeated freezing and thawing.

Moderately frost resistant (M)

Durable except when saturated and subjected to repeated freezing and thawing.

Not frost resistant (O)

Liable to be damaged by freezing and thawing and not normally for use externally. Generally, manufacturers do not make category O bricks intentionally but may classify them as such if for any reason they fail to reach the intended category. They may be suitable for internal use.

Many manufacturers classify frost resistance from the results of laboratory tests as it is seldom practicable to wait for the test of time in use. Although laboratory testing methods have been developed during recent years none was considered appropriate for inclusion in the current British Standard. It is likely that laboratory tests to classify frost resistance will be included in the European Standard for masonry products.

There is no other way of predicting the frost resistance of clay bricks. In particular there is no direct relationship between either compressive strength or water absorption and frost resistance.

Calcium silicate (sandlime and flintlime) bricks

Calcium silicate brick manufacturers are not required, by BS 187[2], to classify their bricks for frost resistance and generally there is little risk of them being damaged by frost. There is a relationship between the compressive strength and frost resistance of calcium silicate bricks and recommendations are made for the use of bricks not less than compressive strength class 4 (27.5 N/mm²) in very exposed conditions.

Figure 6.5. *Frost attack on moderately frost resistant bricks in a capping.*

Mortars – vulnerability at an early age

Mortars are particularly vulnerable before they have set, as once frozen they will not set subsequently. They will be permanently damaged and normally the brickwork will have to be taken down and rebuilt.

Bricklaying should be stopped when a falling air temperature reaches 3°C. Once freezing has occurred bricklaying should not begin again until the temperature reaches 1°C and is rising (see BS 8000[3]), and then providing only that the bricks are not frozen.

Newly built brickwork must be protected from freezing before the mortar has set. *(see Section 1.2 'Protection of newly built brickwork')*

'Antifreeze' admixtures

There are no known 'antifreeze' admixtures that are successful in bricklaying mortars and some are positively harmful. This is referred to in section 4.1 'Mortars' and more fully in section 3.1 'Avoiding damage from extremes of temperature'.

Mortars – long term frost resistance

- Whichever type of mortar is used, cement:lime:sand, cement:sand and plasticiser, or masonry cement:sand, its frost resistance will be enhanced as the proportion of cement is increased. *(see Section 4.1 'Mortars', particularly table 4.1)*
- Where there is a low risk of saturation a designation (iii) mortar will give a good balance of properties for external walling in the UK. In conditions of extreme exposure, either resulting from geographic position or of particular brickwork features, to wind-driven rain and freezing temperatures, designation (ii) or even designation (i) mortars may be advisable partly to provide resistance to sulfate attack. *(see Section 6.3)*
- Designers are responsible for specifying appropriate mortar designations and site supervisors and operatives for accurate batching and correct mixing as described in the section on mortars.
- When pointing brickwork, ensure that the mortar is compacted so that no air pockets remain to fill with water, freeze and dislodge the pointing.

Brickwork features liable to saturation

Brickwork most liable to saturation and freezing includes:

- Horizontal and sloping surfaces, e.g. copings and cappings, sills, projecting courses and plinths.
- Vertical surfaces subject to run-off water from cappings, sills without projections and effective throats, and from areas of glazing and

Figure 6.6. *Frost attack on inadequate designation mortar saturated by water run-off on sloping plinth.*

Figure 6.7. *Frost attack on moderately frost resistant bricks saturated by run-off from hard paving.*

Figure 6.8. *Frost attack on moderately frost resistant bricks not protected from saturation by water from retained earth.*

impervious wall cladding and splashing of rain-water from adjacent hard paving.

- External walls below or within 150 mm above ground level in poorly drained soil.
- Any brickwork in severely exposed areas which may be saturated by driving rain and liable to freezing.

GUIDANCE ON THE CHOICE OF BRICKS AND MORTARS FOR SPECIFIC CONDITIONS

In practice, the selection of suitable bricks and mortars must take into account sulfate attack as well as frost attack. *(see Section 6.3)*

Detailed recommendations for brick and mortar designations are given in table 13 of BS 5628:Part 3[4]. Similar recommendations together with explanatory notes are given in a BDA publication on durability[5].

The recommendations are summarised in this book in section 6.4 'Durability of brickwork'.

In all situations particular care must be taken to batch and mix mortars accurately and effectively.

Figure 6.9. *Moderately frost resistant bricks are satisfactory where protected from saturation by roof eaves.*

KEY POINTS

- Brickwork is at risk from frost attack only when saturated.
- The frost resistance of clay bricks cannot be judged by either their compressive strength or their water absorption.
- Calcium silicate bricks are generally frost resistant and there is a relationship between strength and frost resistance.
- Mortar is particularly vulnerable before it sets.
- Frost resistance of mortars is enhanced with higher proportions of cement.
- Accurate batching and effective mixing is essential.
- 'Antifreeze' admixtures are ineffective in brickwork mortars.

References
(1) BS 3921:1985 Specification for 'Clay bricks'.
(2) BS 187:1978 Specification for 'Calcium silicate (sandlime and flintlime bricks)'.
(3) BS 8000:Part 3:1989 'Workmanship on building sites – Code of practice for masonry' cl. 3.1.1.1.
(4) BS 5628:Part 3:1985 'Use of masonry'.
(5) Brickwork durability' BDA design note 7. September 1986.

6.3   SULFATE ATTACK ON MORTARS

Sulfate attack on mortars is fortunately very rare and slow to develop, but can be costly to remedy. It can be avoided by a few simple safeguards which are mainly the responsibility of designers and specifiers supported by conscientious workmanship. This section provides a necessary understanding of the causes with some recommendations for avoiding and minimising the risk.

THE PROCESS OF SULFATE ALTACK

Sulfate attack on mortars principally results from a chemical reaction between sulfate in solution and a constituent of Portland cement,

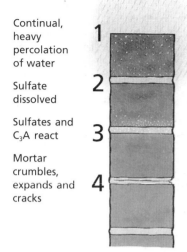

Continual, heavy percolation of water

1

Sulfate dissolved

2

Sulfates and C_3A react

3

Mortar crumbles, expands and cracks

4

Figure 6.10. *Sulfate attack – a simplified diagram.*

Figure 6.11.

Figure 6.12. *Parapet bows from maximum sulfate attack on mortar facing the prevailing wind-driven rain.*

tricalcium aluminate, (C_3A) which forms calcium sulfoaluminate (*ettringite*) *(fig 6.10)*.

The surface of the mortar joint may crumble and erode *(fig 6.11)* and the inside expand, disrupting and even bowing the brickwork *(fig 6.12)*.

The chemical reaction can occur only if ample amounts of water continually percolate through the brickwork and the mortar remains wet for long periods. The slow movement of water by diffusion alone will not carry enough sulfate to the cement.

Water may continually percolate through brickwork in many ways, e.g.

- through copings, cappings or sills that have no effective DPC under them.
- by evaporation through retaining walls in which the retaining faces have no effective damp-proof membranes.

- by running off impervious cladding onto brickwork below or from paving into adjacent brickwork.
- by exceptional exposure to wind-driven rain.

SOURCES OF SULFATE

A high proportion of clay bricks contain some soluble salts, including sulfates. Manufacturers are required by BS 3921[1] to classify their bricks for soluble salts content as either 'Normal' (N) or 'Low' (L) category. The letter specifies limits on particular salts.

Soil, ground or sub-soil waters and made-up ground or fill may contain sulfates which in very wet conditions can penetrate in solution to the cement.

Ashes and clinker, spread over ground to be used for storing materials, can transfer sulfates into bricks stacked in contact with them.

Calcium silicate bricks are virtually free from sulfate and do not contribute to sulfate attack.

AVOIDING AND MINIMISING THE RISK OF SULFATE ATTACK

'Sulphate attack of brickwork will take place only under certain conditions: removal of any one of these conditions would prevent sulphate attack and probably accounts for the relatively few instances of deterioration'[2].

Some of the conditions and ways to eliminate them are referred to here.

Protecting brickwork from saturation

Where a choice exists, the surest way to prevent sulfate attack is to protect the brickwork from being saturated for long periods. This protection can be achieved by careful design and construction that will prevent continual water percolation through the brickwork.

Appropriate and effective projecting and throated copings, sills, verges and eaves, normally protect brickwork from saturation, except possibly in areas of exceptionally severe wind-driven rain.

Brickwork earth-retaining walls should have the retaining face

protected by an effective membrane that has a free-draining material behind.

Provisions where brickwork may become saturated
Cement rich mortars

Sulfate attack can occur only if sulfate solutions penetrate the mortar. Cement-rich mortars, e.g. designations (i) and (ii), are more resistant to water penetration and hence to sulfate despite their higher proportion of C_3A. (see Section 4.1 'Mortars' or Section 6.4 'Durability of brickwork' for the table of mortar designations)

To minimise the risk of brickwork cracking, manufacturers of calcium silicate bricks generally recommend using mortars no stronger than designation (iii). This is normally satisfactory as the bricks are virtually free of soluble salts.

Clay bricks with a 'low' (L) level of soluble salts

Sulfate attack in brickwork built from L category bricks and designation (i) or (ii) mortars is virtually unknown, suggesting that this combination practically eliminates the risk of sulfate attack.

Sulfate-resisting cement

If brickwork is to be built from clay bricks, of 'normal' (N) soluble salts content and it is

likely to remain saturated for long periods, sulfate-resisting cement should be considered in certain conditions. Its use should also be considered where sulfate ground conditions exist.

GUIDANCE ON THE CHOICE OF BRICKS AND MORTARS FOR SPECIFIC CONDITIONS

In practice the selection of suitable bricks and mortars must take into account frost resistance as well as sulfate attack (see Section 6.2).

Detailed recommendations for brick and mortar designations are given in table 13 of BS 5628:Part 3[3].

The recommendations are summarised in this book in section 6.4 'Durability of brickwork'.

References
(1) BS 3921:1985 'Clay bricks'.
(2) Harrison W. H., 'Conditions for sulfate attack on brickwork, chemistry and industry' 19 September 1981.
(3) BS 5628:Part 3:1985 'Use of masonry' (see note below).

NOTE: For those not having access to BS 5628:Part 3, the BDA publication 'Brickwork durability' as revised September 1986, contains tables based on the information in table 13 of the code with useful explanatory notes.

KEY POINTS

- Sulfate attack on mortars is rare but usually costly to remedy.
- It can occur only by persistent water percolation through brickwork.
- Persistent percolation can usually be avoided by appropriate design.

- Experience suggests that in conditions of severe exposure to saturation, if 'low (L)' category bricks are used with cement-rich mortars, sulfate attack is unlikely.
- Sulfate-resisting cement should be considered in certain conditions.

6.4 DURABILITY OF BRICKWORK

The two preceding sections, 6.2 and 6.3, deal with the main causes of the deterioration of brickwork. The more common is frost attack on bricks and mortars. Sulfate attack on mortars is far less common.

Frost and sulfate attack occur only if brickwork becomes saturated. The degree of saturation depends:

- on design details, e.g. projecting and throated sills give better protection than flush sills.
- on workmanship, e.g. badly installed DPCs under copings.
- on the degree of exposure to wind-driven rain, e.g. brickwork facing the prevailing wind across open ground is more exposed than brickwork protected by nearby buildings.

Sulfate attack occurs only if unusually high levels of soluble salts are present, e.g.:

- in clay bricks.
- in some ground conditions.
- in chimney gases.
- in clinker and ashes used on site.

If unusually high levels of soluble salts are present and saturation is unavoidable, the risk of frost and sulfate attack can be minimised by specifying appropriate combinations of:

- brick designations.
- mortar designations.
- sulfate-resisting cement.

Clay brick designations, including frost resistance and soluble salt content categories, are set out in table 6.1.

The classification of engineering and damp-proof course bricks is set out in table 6.2.

Classes of calcium silicate bricks and their compressive strengths are set out in table 6.3.

TABLE 6.1 Durability designations of clay bricks

Designation	Frost resistance (see Section 6.2)	Soluble salts content (see Section 6.3)
FL	Frost resistant (F)	Low (L)
FN	Frost resistant (F)	Normal (N)
ML	Moderately frost resistant (M)	Low (L)
MN	Moderately frost resistant (M)	Normal (N)
OL	Not frost resistant (O)	Low (L)
ON	Not frost resistant (O)	Normal (N)

This table is derived from table 3 of BS 3921[1].

TABLE 6.2 Clay engineering and damp-proof course bricks

Class	Compressive strength N/mm^2	Water absorption % by mass
Engineering A	not less than 70	not more than 4.5
Engineering B	not less than 50	not more than 7.0
DPC 1	not less than 5	not more than 4.5
DPC 2	not less than 5	not more than 7.0

NOTE: There is no direct relationship between compressive strength and water absorption as given in this table and durability.

This table is based on information in table 4 of BS 3921[1].

TABLE 6.3 Calcium silicate (sandlime and flintlime) bricks

Class	Compressive strength N/mm^2
7	48.5
6	41.5
5	34.5
4	27.5
3	20.5

NOTE: There is a strong relationship between the compressive strength of calcium silicate bricks and their durability.

This table is based on information in table 2 of BS 187[2].

TABLE 6.4 Mortar mixes and designations

Basic composition	Cement:lime:sand with air-entrainment		Cement:sand with air-entrainment		Cement:lime:sand
Binders	Ordinary Portland cement or sulfate-resisting Portland cement	Masonry cement with high lime content (1:1 OPC:lime)	Masonry cement with other than high lime content	Ordinary Portland cement or sulfate-resisting Portland cement	
Designation					
(i)					1 : 0–¼:3
(ii)	1 : ½ : 4½ + Air	1 : 3	1 : 2½–3½	1 : 3–4 + Air	1 : ½ : 4–4½
(iii)	1 : 1 : 5½ + Air	1 : 4½	1 : 4–5	1 : 5–6 + Air	1 : 1 : 5–6
(iv)			1 : 5½–6½	1 : 7–8 + Air	1 : 2 : 8–9

▨ – Mortar of high durability   ▬ – General use mortar of good durability

Notes:
The types of mortars of any one designation are of approximately equal strength. The range of sand proportions is to allow for varying grades of sand. The second quantity e.g. 1:1:5–6 for designation (iii) is for a well-graded sand. Smaller proportions of sand (or large proportions of cement and lime) are necessary with less well-graded sands *(see fig 4.3)*.
The proportions of hydrate limes may be increased by up to 50% to improve workability.
With the permission of the designer, plasticisers may be added to lime:sand mixes to improve their early frost resistance. Ready-mixed lime:sand mixes may contain such admixtures. This table is based on information given in Table 15 of BS 5628:Pt3:1985.

TABLE 6.5 Durability of brickwork in finished construction (See Figure 6.13)

Masonry condition or situation	Brick quality/mortar designation				
A Work below or near ground level *(if sulfate ground conditions exist – see note[1])*					
A1 Low risk of saturation with or without freezing	FL	(i)	(ii)	(iii)	
	FN	(i)	(ii)	(iii)	
	ML	(i)	(ii)	(iii)	
	MN	(i)	(ii)	(iii)	
	3–7			(iii)	(iv)[2]
A2 High risk of saturation *without* freezing	FL	(i)	(ii)		
	FN	(i)	(ii)*		
	ML	(i)	(ii)		
	MN	(i)	(ii)*		
	3–7		(ii)	(iii)	
A3 High risk of saturation *with* freezing	FL	(i)	(ii)		
	FN	(i)	(ii)*		
	3–7		(ii)		
B DPCs *(if sulfate ground conditions exist – see note[1])*					
B1 In buildings	DPC1	(i)			
B2 In external works *Note: for classification of DPC bricks see table 6.2.*	DPC2	(i)			

TABLE 6.5 Durability of brickwork in finished construction (continued)

Masonry condition or situation	Brick quality/mortar designation			
C Unrendered external walls *(other than chimneys, copings, cappings, parapets and sills)*				
C1 Low risk of saturation	FL	(i)	(ii)	(iii)
	FN	(i)	(ii)	(iii)
	ML	(i)	(ii)	(iii)
	MN	(i)	(ii)	(iii)
	3–7			(iii) (iv)[2]
C2 High risk of saturation	FL	(i)	(ii)	
	FN	(i)	(ii)*	
	3–7			(iii)
D Rendered external walls *(other than chimneys, copings, cappings, parapets and sills – see note [3])*	FL	(i)	(ii)	(iii)
	FN	(i)*	(ii)	
	ML	(i)	(ii)	(iii)
	MN	(i)*	(ii)	
	3–7			(iii) (iv)[2]
E Internal walls and inner leaves of cavity walls *(where designation (iv) mortars are used – see note [2])*	FL	(i)	(ii)	(iii) (iv)[2]
	FN	(i)	(ii)	(iii) (iv)[2]
	ML	(i)	(ii)	(iii) (iv)[2]
	MN	(i)	(ii)	(iii) (iv)[2]
	OL	(i)	(ii)	(iii) (iv)[2]
	ON	(i)	(ii)	(iii) (iv)[2]
	3–7			(iii) (iv)[2]
F Unrendered parapets *(other than copings and cappings)*				
F1 Low risk of saturation, e.g. low parapets, on some single-storey buildings	FL	(i)	(ii)	(iii)
	FN	(i)	(ii)	(iii)
	ML	(i)	(ii)	(iii)
	MN	(i)	(ii)	(iii)
	3–7			(iii)
F2 High risk of saturation, e.g. where a capping only is provided	FL	(i)	(ii)	
	FN	(i)*	(ii)*	
	3–7			(iii)
G Rendered parapets (other than cappings and copings) *(where sulfate-resisting cement is recommended – see note [3])*	FL	(i)	(ii)	(iii)
	FN	(i)*	(ii)*	
	ML	(i)	(ii)	(iii)
	MN	(i)*	(ii)*	
	3–7			(iii)

TABLE 6.5 Durability of brickwork in finished construction (continued)

Masonry condition or situation	Brick quality/mortar designation			
H Chimneys *(sulfate-resisting cement in mortars and renders are strongly recommended due to the possibility of sulfate attack from flue gases)*				
H1 Unrendered with low risk of saturation	FL FN ML MN 3–7	(i)* (i)* (i)* (i)*	(ii)* (ii)* (ii)* (ii)*	(iii)* (iii)* (iii)* (iii)* (iii)*
H2 Unrendered with high risk of saturation	FL FN 3–7	(i)* (i)*	(ii)* (ii)*	(iii)*
H3 Rendered *(where sulfate-resisting cement is recommended – see note [3])*	FL FN ML MN 3–7	(i)* (i)* (i)* (i)*	(ii)* (ii)* (ii)* (ii)*	(iii)* (iii)* (iii)*
I Cappings, copings and sills [7]				
Cappings, copings and sills except for chimneys	FL FN 4–7	(i) (i)	(ii)	
Cappings and copings for chimneys	FL FN 4–7	(i)* (i)*	(ii)*	
J Freestanding boundary and screen walls *(other than copings and cappings)* J1 With coping [7]	FL FN ML MN 3–7	(i) (i) (i) (i)	(ii) (ii) (ii) (ii)	(iii) (iii) (iii)
J1 (a) With coping exposed to severe driving rain [5] [7]	FL FN ML MN 3–7	(i) (i)* (i) (i)*	(ii) (ii)* (ii) (ii)*	(iii) (iii) (iii)
J2 With capping [7]	FL FN 3–7	(i) (i)	(ii) (ii)	(iii)*[4]

TABLE 6.5 Durability of brickwork in finished construction (continued)				
Masonry condition or situation	Brick quality/mortar designation			
K Earth-retaining walls *(other than copings and cappings)*				
K1 With waterproofed retaining face and coping [7]	FL	(i)	(ii)	
	FN	(i)*	(ii)*	
	ML	(i)	(ii)	
	MN	(i)*	(ii)*	
	3–7		(ii)	(iii)
K2 With coping or capping but no waterproofing on retaining face [7]	FL	(i)		
	FN	(i)*		
	4–7		(ii)	
L Draining and sewerage, e.g. inspection chambers, manholes *(if sulfate ground conditions exist – see note [1])*				
L1 Surface water	Eng	(i)		
	FL	(i)		
	FN	(i)*		
	ML	(i)		
	MN	(i)*		
	3–7		(ii)	(iii)
L2 Foul drainage (continuous contact with bricks)	Eng	(i)		
	FL	(i)		
	FN	(i)*		
	ML	(i)		
	MN	(i)*		
	7[6]		(ii)	
L3 Foul drainage (occasional contact with bricks)	Eng	(i)		
	FL	(i)		
	FN	(i)*		
	ML	(i)		
	MN	(i)*		
	3–7[6]		(ii)	(iii)

KEY:
FL, FN, ML, MN refers to clay brick designations, see table 6.1
3–7 etc. refer to classes of calcium silicate bricks, see table 6.3.
(i), (ii), (iii), (iv) refer to mortar designations, see table 6.4.
* – sulfate-resisting cement is recommended or advisable.

NOTES:
[1]   If sulfate ground conditions exist, expert advice should be taken when specifying mortars. Also see cl. 22.4 BS 5628:part 3.
[2]   Protect masonry under construction from freezing and saturation.
[3]   Where sulfate-resisting cement is recommended for use in mortar it should also be used in the base coat of any rendering.
[4]   Some manufacturers recommend the use of designation (ii) mortars rather than designation (iii) with SRC.
[5]   For notes on assessing exposure to driving rain see section 6.7, *fig 6.46* and text.
[6]   Some types of calcium silicate bricks are not suitable for use in these situations – the manufacturer should be consulted.
[7]   For definitions of copings and cappings – see section 5.1.

This table is based on information given in table 13 of BS 5628:Part 3 (3) which specifiers are advised to refer to as the authoritative document.

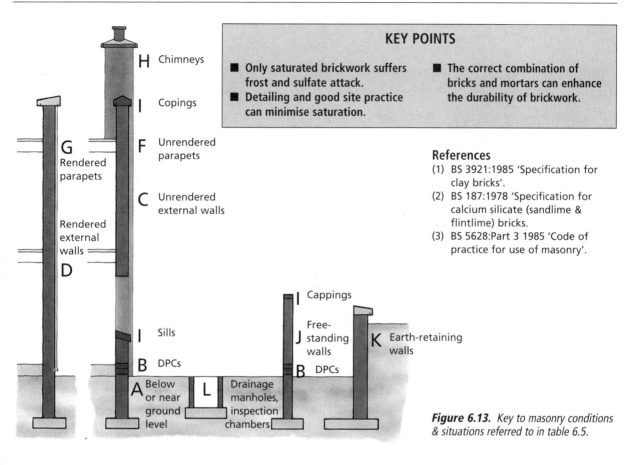

References
(1) BS 3921:1985 'Specification for clay bricks'.
(2) BS 187:1978 'Specification for calcium silicate (sandlime & flintlime) bricks.
(3) BS 5628:Part 3 1985 'Code of practice for use of masonry'.

H Chimneys
I Copings
F Unrendered parapets
G Rendered parapets
C Unrendered external walls
Rendered external walls
D
I Sills
B DPCs
A Below or near ground level
L Drainage manholes, inspection chambers
I Cappings
J Free-standing walls
B DPCs
K Earth-retaining walls

Figure 6.13. Key to masonry conditions & situations referred to in table 6.5.

6.5 ALLOWING FOR VARIATIONS IN BRICK SIZES

The 'actual sizes' of individual bricks within a batch, as well as the average sizes of bricks in different batches, inevitably vary from the 'work sizes'* at which the manufacturers aim (i.e. 215 by 102.5 by 65 mm for standard bricks). This section examines briefly the nature of the variations and ways of allowing for them in producing attractive facework.

Only sufficient background information is given to introduce an essential understanding of the nature of size variation as it affects bricklaying. A more comprehensive exposition of the sophisticated, technical and statistical manufacturing methods used to control variations in size, including the methods of measuring bricks for that purpose, would be inappropriate in this section.

*For definitions see section 2.1 'Setting-out facework' *fig 2.1* and section 6.10 on 'Brick manufacture'.

WHY BRICKS VARY IN SIZE
Despite modern manufacturing and quality control techniques, slight variations in raw materials and firing temperatures, and mould and die wear will cause some variations in sizes within and between batches of bricks, as do some firing techniques deliberately used to produce multicoloured bricks.

HOW BRICKS VARY IN SIZE
Imagine the length of fifty standard bricks, sampled from a

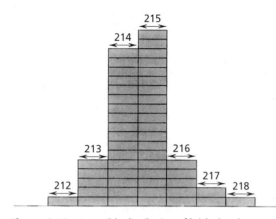

Figure 6.14. *A possible distribution of brick sizes in a sample of fifty.*

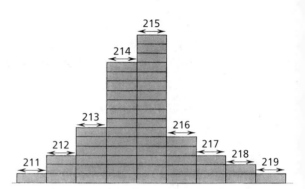

Figure 6.15. *A wider distribution of brick sizes in a sample of fifty.*

consignment, being measured to the nearest millimetre, consistently along the face or centre line. If the bricks of identical length are then stacked in separate piles, with the shortest bricks on the left and the longest on the right the result, with say extruded wire-cut bricks, might be as in *fig 6.14*, while that with hand-made bricks might be as in *fig 6.15*.

In both examples the bricks that are work size or very close to it are in the highest piles, that is to say they occur most frequently, while the large and small sizes at the extremes are in the lowest piles and occur least frequently. *Figure 6.14* depicts a comparatively close distribution of sizes with over three-quarters of the bricks within ± 1 mm of the work size whereas *fig 6.15* depicts a relatively wide distribution with less than half the bricks within ± 1 mm.

The examples are intended to illustrate only the nature of variations in brick sizes and they should not be used to make unwarranted assumptions. For instance, of the fifty bricks, in *fig 6.15*, one is 4 mm more and

another is 4 mm less than work size. It would be wrong to assume from this that 4 per cent of bricks in larger consignments would have a variation of ± 4 mm. Finally, the examples do not illustrate a recognised technique for establishing conformity with the requirements of relevant Standards.

LIMITATIONS ON SIZE VARIATIONS

Limits of sizes for bricks are specified in British Standards, namely BS 187[1] for calcium silicate, BS 3921[2] for fired clay, BS 4729[3] for non-standard (NS) cuboid bricks, BS 6073[4] for concrete and BS 6649[5] for clay

and calcium silicate metric modular bricks. Tolerances for bricks of special shapes and sizes are not specified in BS 4729 but guidance is given in appendix D of that Standard.

HOW BRICKS ARE MEASURED TO CHECK CONFORMITY

Current methods of measuring types of bricks to determine whether they are within the specified limits are shown in *figs 6.16* and *6.17*. It is probable that these methods will be superseded in the European Standard for bricks which is currently being drafted by CEN*, and will, of course, eventually apply to the UK.

Figure 6.16. (inset) *Measuring a single calcium silicate brick.*
Figure 6.17. *Measuring 24 clay bricks.*

But whatever limits and methods of measurement are eventually specified, the need for bricklaying skills to take up variations in brick sizes within the joints will remain a fact of life.

*Comité Européen de Normalisation – translated as European Committee for Standardisation.

HOW ARCHITECTS AND BRICKLAYERS CAN ALLOW FOR VARIATIONS IN BRICK SIZES

Architects should, if possible, dimension brickwork in multiples of whole or half-bricks, to minimise the need for broken bond and cutting of bricks. Multiples of brick dimensions are set out in tabular form for easy reference in BDA Design Note 3[6].

Architects should also identify features, such as narrow piers, soldier courses, copings, cappings and sills where bricks with tighter limits of size than those specified in the relevant British Standard are required. Such bricks may be selected by the manufacturers before delivery or from consignments delivered to site. In both cases instructions should be given well in advance.

It is advisable to warn all the manufacturers where polychrome brickwork is involved so that they can liaise and minimise any effect of variations in sizes between the different colour bricks.

Bricklayers use their traditional skills and experience so that, despite variations in brick sizes, the brickwork will have a uniform appearance with cross joints as consistent as is practicable.

Particular points for both architects and bricklayers are listed below.

Checking deliveries

Take samples of bricks from different packs, from different positions within the packs and in vertical slices. Measure all three dimensions. If it is suspected that the consignment does not comply with the limits of size in the relevant British Standard, first inform the supplier or manufacturer or both. It is advisable to have at least one of them present when carrying out the precise conditions of the test described in the relevant Standard.

Blending packs of bricks

Modern methods of mechanically handling and transporting bricks tend to take bricks from one part of the kiln, and therefore of similar actual sizes, and package them in horizontal layers rather than blend them as happened when bricks were loaded and unloaded by hand. Packaged bricks should be blended on site, when 'loading out' for the bricklayer, by drawing from at least three packs at a time preferably broken down vertically *(fig 6.18). (see Sections 1.3 'Handling, storage and protection of materials'; 3.2 'Blending facing bricks on site')*

Some manufacturers are able to blend bricks before they are packaged, so that extremes of sizes and variations of colour are distributed more evenly. Even so it may still be advisable to 'load out' from three or more packs. If there is any doubt the manufacturer should be asked.

Setting-out at ground level

The first course of brickwork should be 'run-out dry' before bricklaying begins in order to check the size of cross joints. Because the average brick size may vary between batches, facework should be set out to the 'coordinating size', i.e. 4 stretchers to 900 mm rather than attempting to maintain 10 mm joints.

Plan the location of reveal bricks and any broken bond, check bonding arrangements at corners, angles and attached piers particularly if squint or dog leg special shapes or plinth courses are to be built-in and plan the position of perpends. *(see Sections 2.1 'Setting-out facework'; 2.4 'Vertical perpends')*

This procedure is not a sign of a bricklayer who lacks skill or experience but of one who takes care to anticipate and prepare for the work to hand, the better to exercise his skill and avoid costly mistakes.

Figure 6.18. A number of packs opened for 'loading out' to bricklayers.

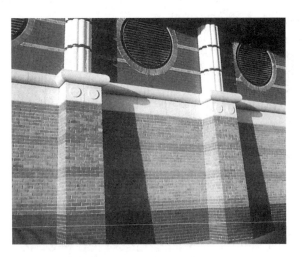

Figure 6.20.
Contrasting coloured
courses at Pumping
Station, Isle of Dogs.
Architect John Outram
Associates.

Figure 6.19. Bands of contrasting coloured bricks at Newnham College, Cambridge. Architect Birken Heywood. (See also inside back cover.)

Contrasting bands of different types of bricks *(figs 6.19 and 6.20)*

If, despite the architect's efforts, bricks for use in contrasting band courses have widely different average lengths from those in the main wall or others in the band, the bricklayer should warn a supervisor of any anticipated difficulty in avoiding excessively wide or narrow vertical cross joints in the bands. Architects may decide to use mortars matching the brick colour to minimise any visual discrepancy.

Excessively large or small bricks

A few may occur in a batch which conforms to the specified limits of size. They should not be discarded but set on one side for possible use elsewhere, e.g. where cut bricks are required for broken bond or at openings.

Figure 6.21. ATRs – not claimed to conform to BS limits of size.

Clay bricks that do not meet the BS limits of size

Some soft-mud, stock bricks, specified for their wide variations in colour and texture are not selected before delivery and may not conform to the British Standard for limits of size. These are often described as ATRs meaning 'as they rise' *(fig 6.21)* and often, though not always, result from clamp firing *(see*

Section 6.10 'Brick manufacture'). Unless a reference panel has been built and agreed, bricklayers should, before beginning to lay such bricks, ensure that the impossibility of maintaining plumb perpends and 10 mm joints is appreciated and that the result will be acceptable. The architect may require some selection to be made on site. *(see Section 1.1 'Reference and sample panels')*

Soldier courses, arches and brick-on-end sills
(figure 6.22)

In order to get a straight line top and bottom, the architect should have arranged for the bricks to be selected in the factory by the manufacturer or by others on site, typically to ± 1.5 mm. Even so the bricklayer may have to set some bricks aside in order to maintain an acceptable line top and bottom.

Cappings and copings
(figures 6.23 and 6.24)

An unbonded coping or capping that has to be aligned both sides demands bricks with very little

Figure 6.22. Soldier courses. River Floss Flood Alleviation Scheme, York. Architects Clouston.

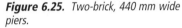

Figure 6.25. Two-brick, 440 mm wide piers.

Figure 6.23. A satisfactory double cant brick-on-edge capping.

Figure 6.24. An unsatisfactory double cant brick-on-edge capping.

Figure 6.26. One-brick, 215 mm wide pier.

size variation and invariably this requires selection, preferably before delivery. Bricklayers may sometimes have to turn double cant bricks end for end to get the best alignment of the cant faces if they are not a perfect 45°. Similar care must be taken when using cant bricks in window sills.

Narrow attached piers

Narrow piers *(fig 6.25)* may require some selection of bricks if the cross joints are to be reasonably uniform in width. One-brick piers *(fig 6.26)* pose special problems as it is possible to plumb only one side of the pier and the other side will vary to the extent of the stretcher lengths. Bricklayers should ask before starting work whether the variation is acceptable or whether selection is required.

References

(1) BS 187:1978 Specification for calcium silicate (sandlime & flintlime) bricks.
(2) BS 3921:1985 Specification for clay bricks.
(3) BS 4729:1990 Specification for dimensions of bricks of special shapes and sizes.
(4) BS 6073:Part 1:1981 Specification for precast masonry units.
(5) BS 6649:1985 Specification for clay and calcium silicate modular bricks.
(6) BDA Design Note 3. Brickwork Dimensions Tables. A guide to designing and building to brick dimensions.

KEY POINTS

- Measure all three faces of a number of bricks from different packs shortly after delivery.
- Inform the supplier or manufacturer immediately if you suspect they do not conform to the specified limits.
- Run-out the first course of bricks dry before starting to build.

- If the building has bands of different colour bricks, run-out dry adjacent rows of each type to compare the relative sizes of the cross joints.
- Check bricks to be used in soldier courses, narrow piers and cappings.
- Draw from at least three packs of bricks when 'loading out'.

6.6 APPEARANCE

The appearance of brickwork depends initially on architects' designs but can be realised only by competent brickwork teams.

THE ARCHITECT'S REQUIREMENTS

The colour, texture, size and shape of the bricks, the colour and profile of the mortar joints and the brickwork bond, all affect appearance. They are normally chosen by the architect.

THE BRICKWORK TEAM'S RESPONSIBILITIES

In fulfilling the design requirements the brickwork team, which may include main and subcontractors, supervisors, bricklayers, mixer drivers and labourers, is responsible for completing many operations requiring skill, knowledge, understanding, care and attention.

Good site management ensures that all members of the team know the operations for which they are responsible.

In this section, comments on sixteen operations are grouped under the main headings that describe those factors that affect the appearance of brickwork.

COLOUR AND TEXTURE OF BRICKS

The choice of colour and texture of bricks is obviously very important.

The brickwork team has important but perhaps less obvious responsibilities for:
1. Checking deliveries
Check that the correct bricks have been delivered, against a reference panel if one has been built *(fig 6.27)*. Reference panels may have been built not only to establish a suitable colour and profile for the mortar joints but also to establish the extent to which colour variations and minor surface blemishes are acceptable. Viewing of the panel for these purposes is usually done from a distance of 3 m. *(see Section 1.1 'Reference and sample panels')*

2. Storage and protection of bricks
Store bricks to avoid contact with the soil, contamination by mud and other building materials, and protect from saturation which can cause staining and efflorescence. *(see Section 1.3 'Handling, storage and protection of materials')*

3. Blending deliveries
Unless bricks have been adequately blended by the manufacturer before delivery the brickwork team must do so by loading out from at least three packs so as to avoid colour patchiness or banding in the brickwork *(fig 6.28)*. *(see Sections 1.3 'Handling, storage and protection of materials'; 3.2 'Blending facing bricks on site')* Some manufacturers recommend loading out from four, five or six packs, especially for bricks having descriptions such as 'Multi', 'Blend' or 'Mixture'.

Particular care should be taken in blending multicoloured bricks which by their nature vary a great deal. In addition, bricklayers, unless instructed otherwise, should select bricks from the stack to ensure that several bricks of a similar colour are not laid adjacent in one course or concentrated in a patch *(fig 6.29)*.

COLOUR OF MORTAR JOINTS
About one fifth the surface area of brickwork consists of mortar joints. As a result the mortar

Figure 6.27. A reference panel.

colour profoundly affects the apparent colour of the brickwork. Used deliberately, as an element of design, changes of mortar joint colour can enhance brickwork *(fig 6.30)*. Unintentional changes caused by lack of skill or carelessness on site can be visually disastrous *(fig 6.31)*.

The colour of mortar depends in the first instance on the colour of the cement, sand and pigments, whether or not lime is used and on the proportions in which these materials are batched. In addition the water content, or consistence, of the mortar and the extent of tooling of the joint which brings the laitence to the surface, all affect the colour.

The bricklaying team should ensure:

4. Consistency of supply of materials for mortars

The colours of sands and even Portland cements from different sources vary enough to affect mortar colours and where colour consistency is required these materials should always be obtained from the same source. Pre-mixed coarse stuff and ready-to-use retarded mortars should also be obtained consistently from the same supplier. *(see Section 4.1 'Mortars')*

5. Protection of mortar materials

Mortar materials and pre-mixed mortars must be protected on site from mud, dirt, oil and other building materials. Protection is also needed against the fine particles of cement, lime and

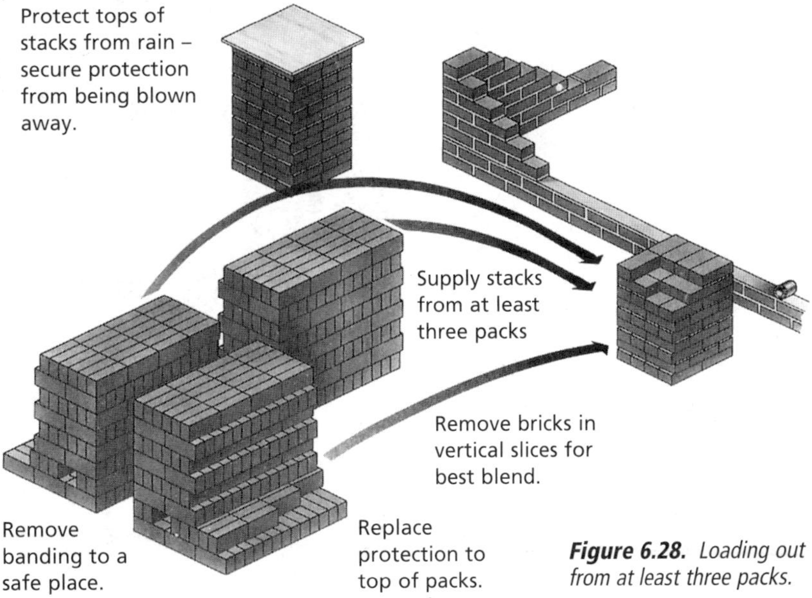

Protect tops of stacks from rain – secure protection from being blown away.

Supply stacks from at least three packs

Remove bricks in vertical slices for best blend.

Remove banding to a safe place.

Replace protection to top of packs.

Figure 6.28. *Loading out from at least three packs.*

Figure 6.29. *Multicoloured bricks – dispersing bricks of similar colour.*

Figure 6.30. *Intentional decorative use of coloured mortars. (See also inside cover.)*

Figure 6.31. *Unintentional changes of mortar colour.*

pigments being washed out by water. Lack of such protection can result in mortar colour changes. *(see Section 1.3 'Handling, storage and protection of materials')*

6. Accurate and consistent proportioning and mixing
Whether separate materials are proportioned on site, by weight or volume, or whether cement is added to pre-mixed coarse stuff, accuracy must be maintained *(fig 6.32)*, otherwise batches of mortar will differ in colour. Always ask the mortar supplier before adding anything which might affect the colour of the mortar. It may also affect the strength.

Similarly mortar must be mixed in a consistent way whether by hand or machine in order to avoid colour variations. *(see Section 4.1 'Mortars')*

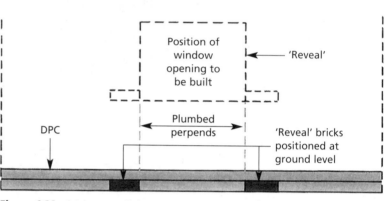

Figure 6.32.
Purpose-made boxes for accurate gauging of mortar materials.

Figure 6.33. Setting-out window position at ground level.

QUALITY OF FACEWORK
Quality brickwork is distinguished from that which is merely adequate by the care with which facework is set-out to minimise the effect of unavoidable broken bond, to maintain vertical perpends, line, level and plumb and regular cross and bed joint widths. The matters dealt with under this heading are more fully described in other sections. *(see Sections 2.1 'Setting-out facework – stretcher half-bond'; 2.2 'Gauge and storey rods'; 2.3 'Line, level and plumb'; 2.4 'Vertical perpends')*

The brickwork team should:

7. Set-out dry at ground level
Set-out brickwork dry at ground level, noting the position of

reveals to openings above *(fig 6.33)*. Consult the supervisor or architect to agree the position of any unavoidable broken or reverse bond *(fig 6.34)*. *(see Section 2.1 'Setting-out facework – stretcher half-bond')*

Establish perpends at ground level and plumb at regular intervals for the full height of facework to avoid the visual distraction of their wandering across the facade *(fig 6.35)*. *(see Section 2.4 'Vertical perpends')*

8. Lay to the line
Bricklaying line should be fine enough for the top arris of every brick to 'follow the line'. Each brick should be close without touching, so that you can 'just see the light' between brick and

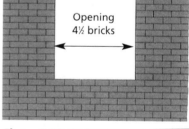

Figure 6.34. Broken bond positioned under window.

line *(fig 6.36) (see Section 2.3 Line, level and plumb')*. When laying some stock type bricks it may be necessary to modify the normal technique and allow some portions of irregular brick faces to be in front of the line in order to obtain a generally regular appearance.

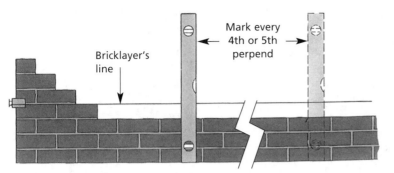

Figure 6.35. *Plumbing perpends at intervals at ground level.*

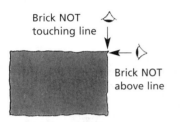

Figure 6.36. *Laying to a line.*

Figure 6.37. *'Hatching and grinning'.*

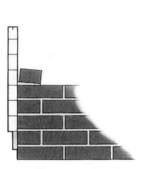

Figure 6.38. *Checking gauge.*

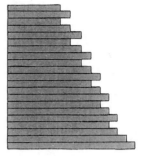

Building corners –
unsatisfactory method

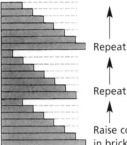

Building corners –
recommended method

Figure 6.39.

9. Avoid 'hatching and grinning'

The lower arris of each brick must be bedded flush with the course below. If bricks are laid carelessly so that the lower arrises are not flush with those below, 'hatching and grinning' will be apparent, particularly when the sun shines obliquely down the face *(fig 6.37)*.

10. Maintain vertical gauge

When raising quoins, maintain consistent vertical gauging of bed joints, by the use of gauge and storey rods *(fig 6.38)*. *(see section 2.2 'Gauge and storey rods')*

11. Rack back wherever possible

Rack back incomplete work and avoid vertical toothing wherever possible. It is difficult to build on to toothing without the join showing on the facework *(fig 6.39)*. *(see Section 2.3 'Line, level and plumb')*

12. Lay some stock bricks with extra care

When laying some soft mud or stock bricks a different technique is required from that used when laying wire-cut or pressed bricks.

The upper and lower surfaces of the latter are generally at right angles to their stretcher and header faces. This will not always be the case with stock type bricks especially if they are clamp fired.

When laying slightly distorted bricks to the line, tilt them one way or the other in order to get the brick face into the same vertical face plane as the wall *(fig 6.40)*. This means that the top of the brick cannot be levelled across the wall.

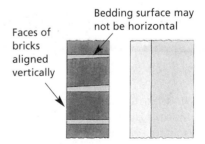

Figure 6.40. *Laying distorted stock bricks.*

Figure 6.41. *Laying a hand-made brick frog up with a 'smile' on its face.*

Figure 6.42. *Finishing of joints must be consistent.*

Figure 6.43. *Protecting newly built brickwork.*

Figure 6.44. *Protecting vulnerable brickwork.*

Figure 6.45. *Plinth, protected from mortar droppings.*

The resulting irregularity of the top surface will be taken up in the bedding mortar and be unnoticed in the finished wall.

Failure to modify normal techniques when necessary can cause an uneven surface known as 'hatching and grinning'. *(see point 9.)*

Unless otherwise instructed, lay hand-made bricks consistently with the frog up so that the creases read as a 'smile' on the face *(fig 6.41)*.

13. Work cleanly

Avoid smudging the face of bricks by spreading just the right amount of mortar each time and cleanly cutting off any surplus. Clean off any splashes of mortar and turn back scaffold boards next to the brickwork at the end of the day to avoid mortar being splashed by rainfall onto the facework.

When sweeping or removing scaffold boards take care not to spill mortar on the completed facework.

14. Maintain a consistent jointing technique

Whatever joint profile and finish is specified, the techniques of all bricklayers must be co-ordinated to provide consistency in the type and diameter and shape of jointer, the pressure used, the angle of striking and depth of recessing, otherwise patches or bands of apparently different coloured bnckwork will be evident *(fig 6.42)*.

Take care that struck perpends are struck on the same side, normally the left-hand side of all cross joints is indented for right and left handed bricklayers alike. *(see Sections 2.7 'Finishing mortar joints'; 2.8 'Pointing and repointing'; 4.1 'Mortars')*

15. Protect newly built brickwork

Rain on newly built brickwork can saturate the mortar and wash out fine particles of cement, lime and pigments, changing the colour of the mortar. Worse still, saturation of the brickwork can lead to severe lime-staining on both the mortar joints and the brick faces. Provision should be made to protect newly built brickwork *(fig 6.43)*. *(see Section 1.2 'Protection of newly built brickwork')*

Vulnerable corners and reveals should be protected from impact damage *(fig 6.44)*. Brickwork features such as projecting plinths should be protected from mortar droppings and damage from falling objects *(fig 6.45)*.

SEALANTS TO MOVEMENT JOINTS

16. A wide choice of sealant colours is available to blend with mortar colours. Ensure that the sealant used conforms to the specification but if none exists that approval is obtained before any sealant is applied.

KEY POINTS	
■ Check deliveries of bricks for colour and texture.	■ Set-out bond dry at ground level.
■ Blend bricks by loading out from at least three packs.	■ Determine positions of openings, any broken bond and perpends.
■ Ensure consistency of source for pre-mixed mortars and mortar materials.	■ Maintain line, level, plumb and gauge.
■ Protect mortar materials and pre-mixed mortars.	■ Keep facework clear of mortar stains.
■ Accurately proportion and consistently mix mortars.	■ Protect newly built brickwork.

6.7   RAIN RESISTANCE OF CAVITY WALLS

THE BRICKLAYER'S CONTRIBUTION

One of the most common building defects is rain penetration causing damage to internal finishes and other vulnerable materials. It usually results from inadequate design, workmanship or both. Remedies are invariably expensive and disruptive.

Bricklayers with knowledge, care and skill can maximise the resistance to wind-driven rain of masonry cavity walls.

Cavity walls began to supersede solid walls in the 1930s because, when properly designed and built, they were more resistant to rain penetration. But since the 1970s, the use of cavity insulation has increased the risk of rain penetration and the need for careful design and workmanship.

DESIGN
Assessing exposure to wind-driven rain
Designers assess the severity and frequency of wind-driven rain which a particular wall will have

to resist, by reference to BS 8104:1992[1]. The assessment takes into account:
a) the geographic location of the wall within the UK *(fig 6.46a)*.

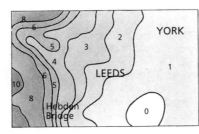

a. Geographic location.

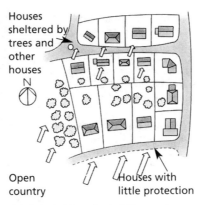

c. Sheltered by other buildings and trees.

b) the formation of nearby land, including marked changes of level *(fig 6.46b)*.
c) other buildings and trees which may shelter the wall; the orientation of the wall *(fig 6.46c)*.

b. Land formation.

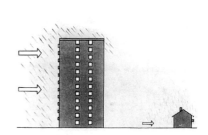

d. Design of walls, e.g. height and overhanging eaves.

Figure 6.46. *Factors considered when assessing exposure to wind-driven rain.*
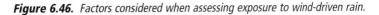

d) the design of the wall, including height, length and protective overhangs (fig 6.46d).

Detailed design and specification

When designing details and specifying materials to meet the assessed exposure, designers refer to BS 5628:Part 3[2], which classifies the exposure of sites relative to the severity of exposure and gives guidance on the factors that affect resistance to rain penetration.

But even the best designs depend on bricklaying skills, care, knowledge and an understanding of those **aspects requiring special care**.

WORKMANSHIP
Aspects requiring special care
(fig 6.47a)

Bricklayers have a responsibility to:

1) **Maximise** the rain resistance of the **outer leaf**.
2) Ensure that **wall ties** do not transmit water across the cavity.
3) Build-in cavity **trays**, vertical DPCs and form **weep holes** to intercept water reaching the inner face of the outer leaf and drain it to the outside.
4) Build-in **thermal insulation** so that it does not provide a path for water penetration across the cavity.

Further detailed requirements for care when building cavity walls are shown in *fig 6.47b* and described in the rest of this section.

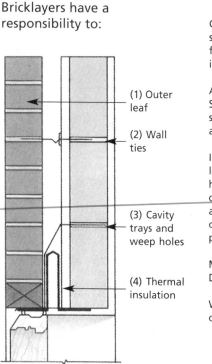

(1) Outer leaf

(2) Wall ties

(3) Cavity trays and weep holes

(4) Thermal insulation

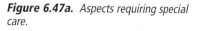

Figure 6.47a. *Aspects requiring special care.*

With full-fill or no cavity insulation – a minimum 50 mm cavity between leaves

With partial-fill cavity insulation – a recommended minimum 50 mm residual air space

Clean excess mortar from cavity side of both leaves, especially from outer leaf when building-in full-fill insulation

Avoid protrusions in cavity. Snapped headers, if required, should be purpose-made or accurately and cleanly cut

Immediately above DPC trays leave cross joints open as weep holes at not more than 1 m centres but with at least two above any opening. Keep them clear of debris. Fit filtration plugs if required

Minimum 150 mm between DPC and ground level

Weep holes every fourth cross joint

Suspend lath to minimise mortar falling down cavity. Remove and clean after six courses. 'Ropes' of twisted hessian, about 3 m long, may be positioned above trays and periodically carefully drawn out through coring holes

Clean mortar droppings from ties and cavity trays as work proceeds. Do not damage trays

Step cavity tray up to inner leaf at least 150 mm

Minimum 150 mm between DPC and bottom of cavity

Leave shallow space at ground level for unavoidable mortar droppings

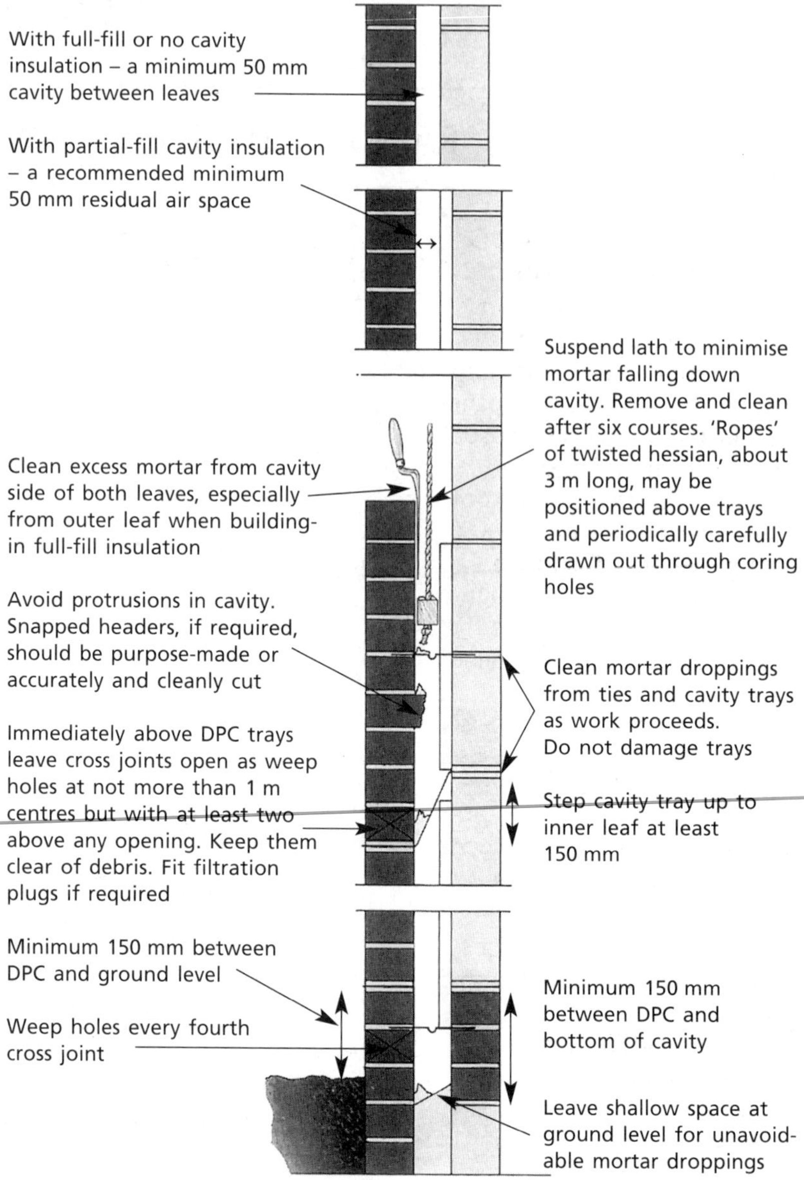

Figure 6.47b. *Further requirements for care when building cavity walls.*

1. MAXIMISE RAIN RESISTANCE OF THE OUTER LEAF

'. . .some water will inevitably penetrate the outer masonry leaf in long periods of wind-driven rain. . .'[3]. The quantity and degree of penetration depend largely on the intensity and duration of wind and rain.

During light, wind-driven rain, damp patches usually appear first at the joints on the cavity face *(fig 6.48a)*. When the rain stops they dry out. After longer or more intense periods of wind-driven rain, the entire face may become wet and eventually water may run freely down the face *(figs 6.48b and c)*.

Penetration of a leaf built from low absorption bricks will generally occur more quickly than through one built from high absorption bricks in the same exposure conditions. The latter delays penetration but it will occur eventually if the wind-driven rain continues long enough for the bricks to become saturated. This is often referred to as the raincoat and overcoat effect' *(figs 6.48d and e)*.

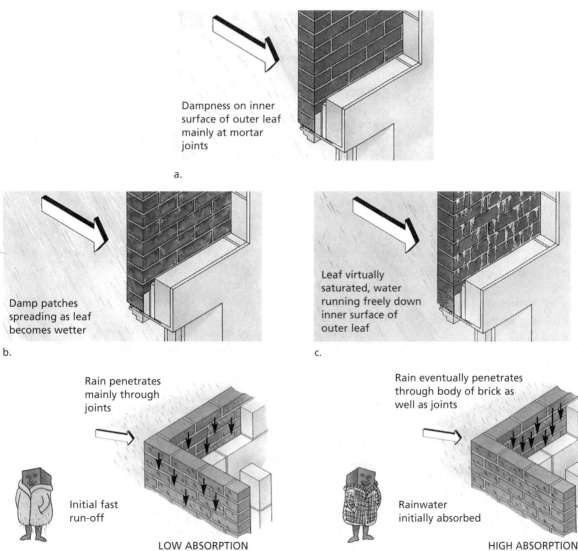

Dampness on inner surface of outer leaf mainly at mortar joints

a.

Damp patches spreading as leaf becomes wetter

b.

Leaf virtually saturated, water running freely down inner surface of outer leaf

c.

Rain penetrates mainly through joints

Initial fast run-off

LOW ABSORPTION BRICKS

d. The 'Raincoat' effect

Rain eventually penetrates through body of brick as well as joints

Rainwater initially absorbed

HIGH ABSORPTION BRICKS

e. The 'Overcoat' effect

Figure 6.48. *Stages of rain penetration of outer leaf under increasing conditions of exposure to wind-driven rain.*

Mortar joints are vulnerable.
Tests and observations show that rainwater first penetrates outer leaves in two ways, through badly filled mortar joints and fine cracks at the brick/mortar interface.

i) **Partially filled joints.** Fill all joints completely. Do not leave a joint hollow by merely 'tipping and tailing' the brick end *(fig 6.49)*. *Remember, there are sixty cross joints to every square metre of wall!*

ii) **Brick/mortar interfaces.** Once bricks are bedded, do not attempt to adjust their position. *This can break the bond, leaving fine cracks for rain penetration at brick/mortar interfaces.*

iii) **Joint profiles.** Form joint profiles as instructed *(see Section 6.9 'Bricklaying tools and equipment')*. *Mortar joints that are finished by tooling with the point of a trowel or a special jointing iron (e.g. struck or weathered and bucket handle) have the surface compressed and the mortar pressed into intimate contact with the bricks to give maximum rain resistance (fig 6.50).* Recessed

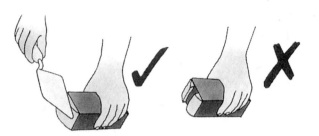

Figure 6.49. Fill – do not 'tip and tail' cross joints.

joints allow more rainwater to penetrate the outer leaf than joints with other profiles and should be specified only in sheltered positions.

iv) **Mortar mixes.** Use the specified mortar mix; in particular do not omit lime if it is specified. *Lime, being water retentive, delays drying of the mortar, allowing time for a good adhesion to take place which is considered to improve rain resistance.* *(see Section 4.1 'Mortars')*

2. WALL TIES

Do not slope wall ties down to the inner leaf; position drips in the centre of the cavity pointing down; keep ties free of mortar droppings *(fig 6.51)*. *Wall ties can become a major cause of rain transfer across the cavity: in any square metre of wall*

there will be at least 5 ties, each a potential path for rain penetration unless built-in properly. (see Section 4.2 'Ties in cavity walls')

3. CAVITY TRAYS, DPCS AND WEEP HOLES

(See also Section 4.3 'Damp-proof courses')
Cavity trays should be positioned immediately above anything which bridges a cavity, such as lintels, support angles and floor beams (but not wall ties). Their purpose is to drain the water to the outside via weep holes *(fig 6.52)*.

Bed cavity trays preferably in a single length and extend them up the inner leaf by at least 150 mm. Bed the top edge into a mortar bed joint or fix flush with the inner leaf in accordance with the manufacturer's instructions. Lap unavoidable

Figure 6.50. Tooling a 'bucket handle' profile.

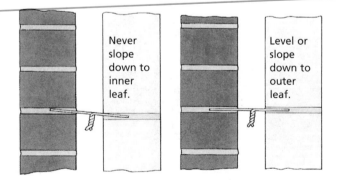

Never slope down to inner leaf.

Level or slope down to outer leaf.

Figure 6.51. Positioning wall ties.

joints in straight lengths, at corners, changes of levels and stop ends by a minimum of 100 mm and bond them with an adhesive recommended by the manufacturers. *The use of preformed units at corners and changes in level avoids the need for complex cutting and folding, which is particularly difficult on site. Many instances of rain penetration are traced to laps which have not been effectively sealed.*

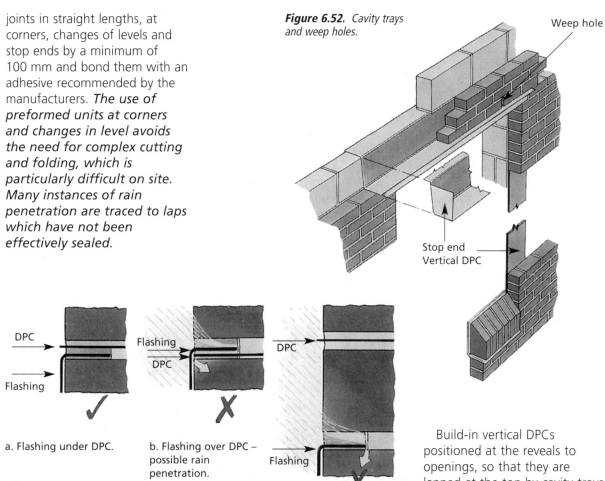

Figure 6.52. *Cavity trays and weep holes.*

Weep hole

Stop end
Vertical DPC

a. Flashing under DPC.

b. Flashing over DPC – possible rain penetration.

DPC

Flashing

DPC

DPC

Flashing

Flashing

Figure 6.53. *Right and wrong ways of positioning DPCs and flashings.*

Bed DPCs and cavity trays on mortar to provide a good bond and project them 5 mm beyond or leave flush with the face of the brickwork. Never recess and cover DPCs with mortar otherwise the surface of the brick and mortar joint may spall as the DPC compresses under load. Never allow ground-level DPCs to project into the cavities, as mortar can build up and bridge the cavity.

Securely bond specified stop-ends to cavity trays to prevent water running into cavities and any insulation that may be present or introduced at a later date. If no stop ends are supplied ask a supervisor if this is correct. if there is to be no cavity insulation you may be instructed not to fit stop ends but in this case cavity trays should be extended 150 mm beyond the end of lintels.

Form weep holes by leaving open cross joints at 1 m centres maximum. Generally, there should be no less than two over any opening. If required insert or build-in proprietary plastic formers or fibrous filters at the face of the joints.

Build-in vertical DPCs positioned at the reveals to openings, so that they are lapped at the top by cavity trays and in turn lap sill DPCs at the bottom. Where cavities are closed project the DPC into the cavity by at least 23 mm but preferably 50 mm. These matters have been dealt with in sections 3.3 'External cavity walls'; 4.3 'Damp-proof courses'.

When building cavity parapet walls take particular care that the flashings which make-good the joint between the tray and the roof finish are correctly installed *(fig 6.53). Failure to provide a satisfactory junction between flashings and DPCs is another common cause of rain penetration. (see Section 5.2 'Cavity parapet walls')*

4. THERMAL INSULATION

Fix all cavity batts, slabs or boards in accordance with the manufacturer's instructions and in particular butt the edges closely, but do not overlap. Ensure that they are kept free from mortar droppings which could bridge the cavity and form a path for rain penetration.

Place a temporary protective board on the top edges of the insulation to prevent mortar droppings collecting and forming a bridge across the cavity. When ready to continue, remove the temporary protection and clean off any mortar before placing the next insulation board, slab or batt *(fig 6.54)*. Do not build-in damaged or wet insulation.

Fix partial-fill insulation slabs or boards securely with proper fittings, usually to the inner leaf, so that no part can tilt across the cavity and form a ledge for mortar droppings. Do not use the drips of butterfly ties for this purpose; they are not large enough. Generally, retain a 50 mm cavity in front of the insulation. Sometimes, in sheltered areas, 25 mm may be specified but this requires great care by the bricklayer if the cavity is not to be bridged.

With full-fill cavity batts take particular care to clean excess mortar from joints on the cavity side of the outer leaf. This is easier to do if the outer leaf is built first. *Hardened mortar protruding into the horizontal joint between batts provides a path for water penetration (fig 6.55).*

For further details including avoiding cut edges touching the outer leaf, fitting batts over ties which do not coincide with joints between batts, (*see Section 4.4 'Insulated cavity walls'*).

FINALLY –

Waterproof finishes

Reliance should not be placed on clear 'waterproof' coatings as a substitute for good design and workmanship, even though in some cases their use may reduce rain penetration for a limited period. There is often a risk that by reducing the rate of drying to the outside, the build-up of

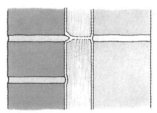

Figure 6.55. Excess mortar causes rain penetration at horizontal joint in insulation.

water behind the outer face may cause further deterioration such as frost attack or sulfate attack on the mortar.

Never apply such finishes to an external leaf of 'M' clay bricks if the cavity is filled with insulation. *The external finish and the insulation both greatly reduce the drying rate of any water entering the inner leaf, which greatly increases the risk of brickwork deterioration.*

References
(1) BS 8104:1992 'Assessing exposure of walls to wind-driven rain'.
(2) BS 5628:Part 3:1985 'Code of practice for masonry'.
(3) Ibid clause 21.1.

Figure 6.54. Protecting insulation from mortar droppings with a temporary board.

KEY POINTS

- Fill all mortar joints solidly.
- Do not adjust brickwork after bedding bricks.
- Form profile of mortar joint as specified.
- Use only the type of DPCs specified.
- Bed all DPCs on fresh mortar.
- Project DPCs 5 mm in front of, or flush with, brickwork face.
- Do not allow DPC to project into cavity.
- Step DPC tray at least 150 mm up the inner leaf; build-in or fix top edge.
- Leave open cross joints as weep holes at not more than 1 m centres.

- Lap flashings and DPCs correctly at the junction of parapet walls and roofs.
- Build-in wall ties level or sloping down to the outer leaf.
- Fix partial-fill slabs or boards to inner leaf with special clips.
- Keep ties free of mortar droppings.
- Keep cavities free of mortar droppings.
- Close butt insulation batts and boards and keep joints free of mortar.
- Clean excess mortar from mortar joints within the cavity.
- Place a temporary protective board over insulation when raising walls above.

6.8   READING CONSTRUCTION DRAWINGS

Bricklayers, as skilled and responsible construction workers, must be able to read drawings accurately. This section suggests some first steps towards the development of this essential skill. Trainees should also take every opportunity to study drawings and wherever possible compare them with a building under construction. Making drawings with a drawing board, T square and scale rule is the best way for trainees to become fluent. That is the way architects learn.

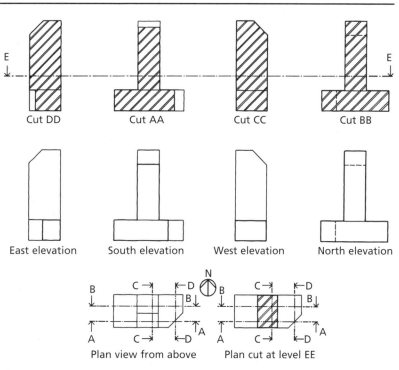

Cut DD   Cut AA   Cut CC   Cut BB

East elevation   South elevation   West elevation   North elevation

Plan view from above   Plan cut at level EE

INTRODUCTORY EXERCISE IN INTERPRETING TWO-DIMENSIONAL DRAWINGS

Carefully study *fig 6.56* and position appropriate bricks dry. Then try the more difficult *fig 6.57*. Check the solutions against *fig 6.60* on page 199. Devise similar simple constructions, make drawings and check with a tutor.

Figure 6.56. *Assembly of two single-cant bricks drawn to a scale of 1:10.*

Figure 6.57. *Assembly of six standard bricks drawn to scale of 1:10.*

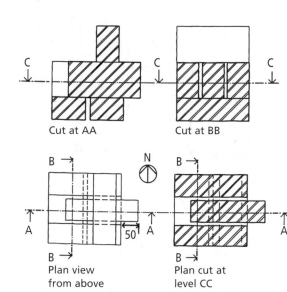

Cut at AA   Cut at BB

Plan view from above

Plan cut at level CC

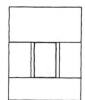

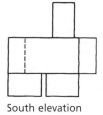

East elevation   South elevation

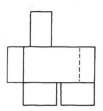

West elevation   North elevation

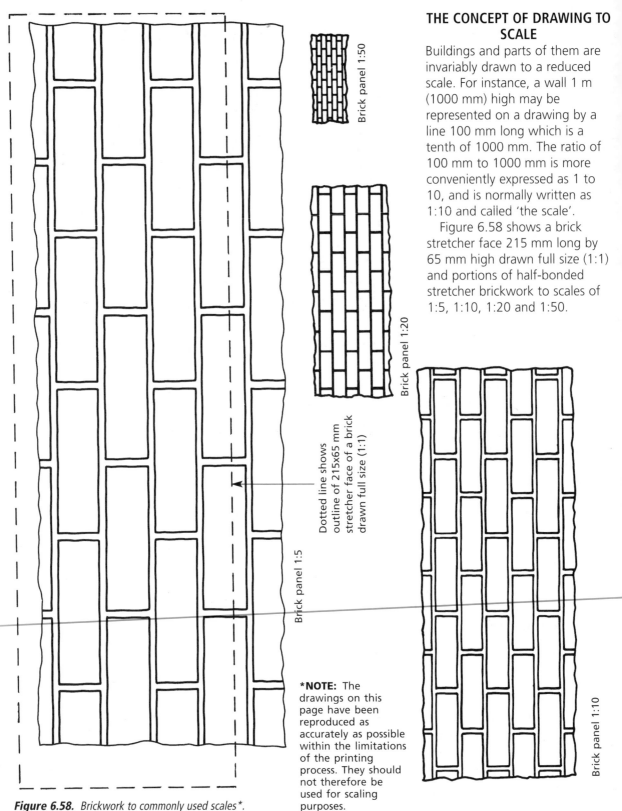

Figure 6.58. *Brickwork to commonly used scales*.*

Brick panel 1:50

Brick panel 1:20

Brick panel 1:5

Dotted line shows outline of 215x65 mm stretcher face of a brick drawn full size (1:1)

Brick panel 1:10

THE CONCEPT OF DRAWING TO SCALE

Buildings and parts of them are invariably drawn to a reduced scale. For instance, a wall 1 m (1000 mm) high may be represented on a drawing by a line 100 mm long which is a tenth of 1000 mm. The ratio of 100 mm to 1000 mm is more conveniently expressed as 1 to 10, and is normally written as 1:10 and called 'the scale'.

Figure 6.58 shows a brick stretcher face 215 mm long by 65 mm high drawn full size (1:1) and portions of half-bonded stretcher brickwork to scales of 1:5, 1:10, 1:20 and 1:50.

*NOTE: The drawings on this page have been reproduced as accurately as possible within the limitations of the printing process. They should not therefore be used for scaling purposes.

Figure 6.59 shows a gable wall of the house in *figs 6.65 & 6.66* to illustrate five smaller scales. A scale of 1:100 is commonly used to show the layout of spaces or rooms, particularly within small buildings like houses. A scale of 1:200 is often more convenient to show the layout of larger buildings. A scale of 1:500 is commonly used for site plans *(fig 6.64)* which show just the outline of buildings in relation to the boundaries of the building site. A logical progression of scales would include 1:1000 and 1:2000 but in practice 1:1250 and 1:2500 are more commonly used for block plans *(fig 6.63)* as they can be based on British Ordnance Survey maps to these scales.

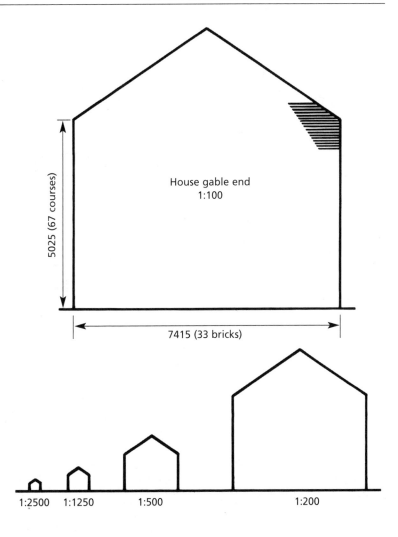

House gable end
1:100

5025 (67 courses)

7415 (33 bricks)

Figure 6.59. *Typical building in commonly used scales.*

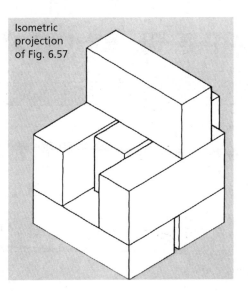

1:2500   1:1250   1:500   1:200

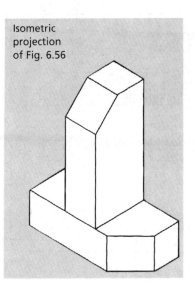

Isometric projection of Fig. 6.56

Isometric projection of Fig. 6.57

Figure 6.60. *The solution of the exercise on page 197.*

USING SCALE RULES

Figure 6.61 illustrates portions of a typical scale rule used to set out dimensions on a drawing. Scale rules are usually marked with eight different scales, two on each of four edges using both sides of the rule.

The dimensions engraved along the edges of scale rules are those represented by the scale rule and are not the actual dimensions as are those marked along the edge of steel tapes and 'school rulers'. For example on the 1:5 scale the numbered dimensions 100, 200 and 300 mm are marked at distances of 20, 40 and 60 mm along the edge of the rule. Note how a distance of 100 mm can represent 100 mm, 500 mm, 2 m, 5 m, 10 m, 20 m, 125 m or 250 m.

A special brickwork scale rule has been developed by the Guild of Bricklayers *(fig 6.62)*. On one side is set out the length and height of bricks to scales of 1:20 and 1:10 which facilitates designing and drawing to brickwork dimensions.

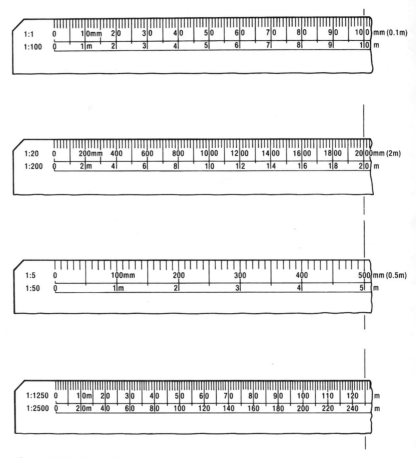

Figure 6.61. *Eight ratios commonly used on scale rules.*

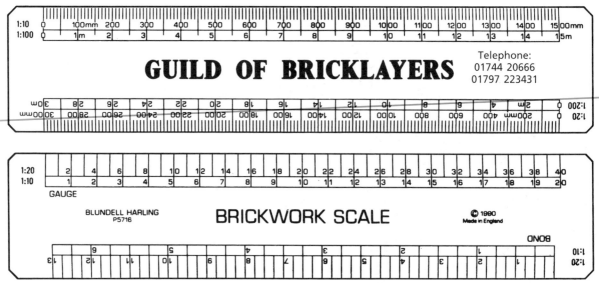

Figure 6.62. *A special brickwork scale.*

REPRESENTING THE LOCATION AND FORM OF A HOUSE BY DRAWINGS

Block plans

These show the location of the building site in relation to adjacent sites, roads and other features *(fig 6.63)*. They are generally drawn to the scales of 1:1250 or 1:2500 which are used for Ordnance Survey maps, although increasingly a more logical sequence of scales, 1:1000 and 1:2000, is being used.

Site plans

These show the location of buildings and possibly drainage, roads, paths and landscaping in relation to the boundaries of the building site and setting-out points and are usually drawn at a scale of 1:500 *(fig 6.64)*.

Layout drawings

These show the spaces or rooms within a building using plans, elevations, sections and cuts.

Plans and elevations

The outside of a house is represented in *fig 6.65*, firstly by two drawings in isometric projection. Both are views from above and from the opposing NW and SE corners. Architects sometimes produce isometric or other 'three-dimensional' drawings but more usually provide only 'two-dimensional' plans (viewed from above) and elevations (facades viewed at right angles).

Figure 6.63. *Block plan 1:2500.*

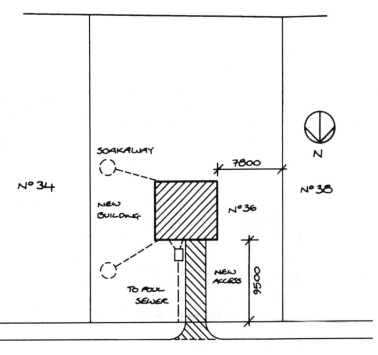

Figure 6.64. *Site plan 1:500.*

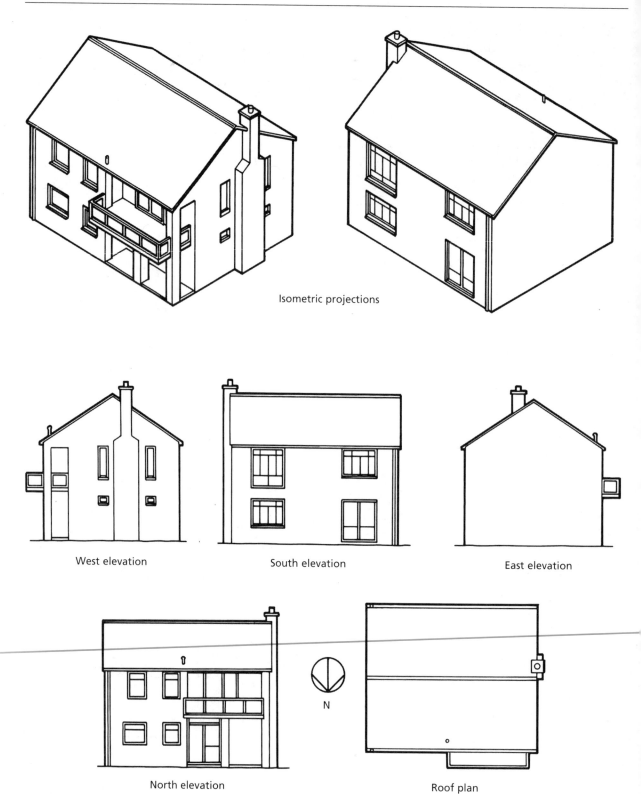

Isometric projections

West elevation

South elevation

East elevation

North elevation

N

Roof plan

Figure 6.65. House exterior represented in three dimensions and by plan and elevations.

Plan cuts and sections*

Imagine that the roof and ceiling are lifted off the walls of the house in *fig 6.65*. It would look somewhat like *fig 6.66a*, but would be drawn as a First Floor Plan *(fig 6.66b)*. If the first floor and ceiling are now removed it would look somewhat like *fig 6.66c* but be drawn as a Ground Floor Plan *(fig 6.66d)*.

Vertical cuts and sections

Cutting vertically through the house *(fig 6.66e)* reveals the roof

and floor structures and the partitions which would be drawn as a vertical cut *(fig 6.66f)*. The position at which a vertical cut or section is taken is shown on the plans. The position at which a horizontal or plan cut or section is taken is shown on the vertical section, often giving the height above a datum level fixed on site.

NOTE: A 'section' shows **only** those features that are on a plane cut through the building. A 'cut' shows, in addition, features beyond the cutting plane. This note is derived from the definitions in BS 1192:Part 1[(1)].

NOTE TO TUTORS: The house design chosen contains a number of questionable features to provide tutors with a means of encouraging the examination and discussion of the design as a way of developing the ability to 'walk through a building' and visualise spaces and construction. For example, what unusual features does the house have? The living room is on the first floor. Describe the route from the front door to the living room and from the ground floor bedroom to the bathroom. There is also one deliberate error in the dimensions in Figure 6.66d.

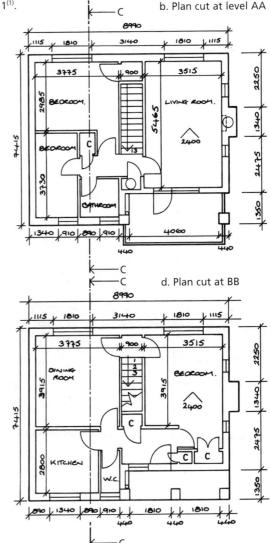

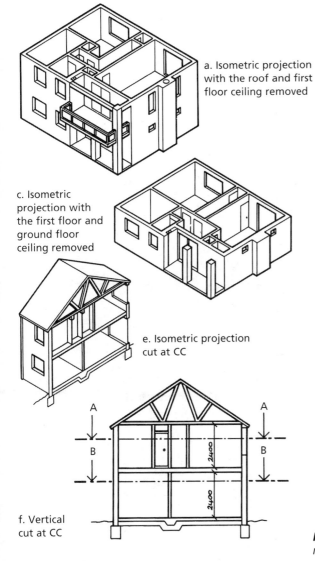

a. Isometric projection with the roof and first floor ceiling removed

c. Isometric projection with the first floor and ground floor ceiling removed

e. Isometric projection cut at CC

f. Vertical cut at CC

b. Plan cut at level AA

d. Plan cut at BB

Figure 6.66. *Plan and vertical sections related to 3-dimensional representations by isometric projections.*

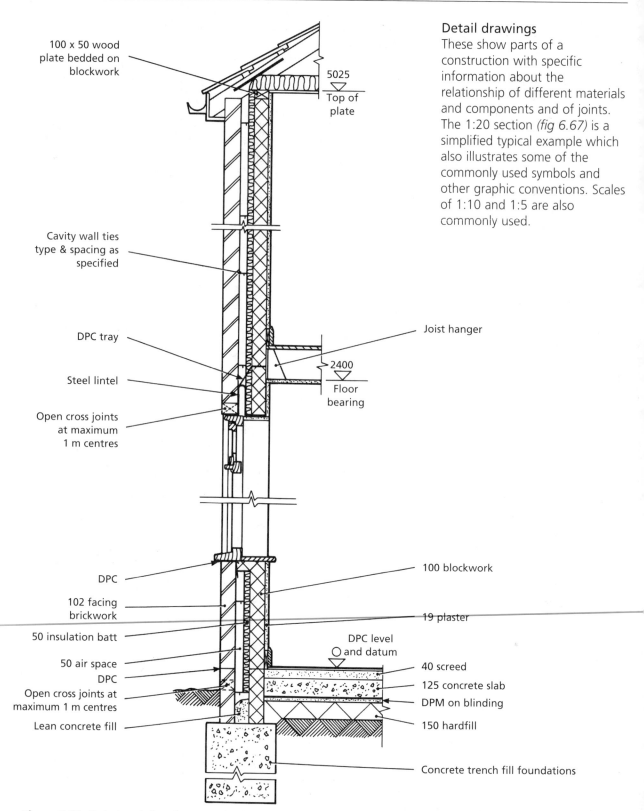

100 x 50 wood plate bedded on blockwork

5025 ▽ Top of plate

Cavity wall ties type & spacing as specified

Joist hanger

DPC tray

2400 ▽ Floor bearing

Steel lintel

Open cross joints at maximum 1 m centres

DPC

100 blockwork

102 facing brickwork

19 plaster

50 insulation batt

DPC level ○ and datum ▽

50 air space

40 screed

DPC

125 concrete slab

Open cross joints at maximum 1 m centres

DPM on blinding

Lean concrete fill

150 hardfill

Concrete trench fill foundations

Detail drawings

These show parts of a construction with specific information about the relationship of different materials and components and of joints. The 1:20 section *(fig 6.67)* is a simplified typical example which also illustrates some of the commonly used symbols and other graphic conventions. Scales of 1:10 and 1:5 are also commonly used.

Figure 6.67. *Typical vertical section at a scale of 1:20.*

CONVENTIONS AND SYMBOLS USED ON DRAWINGS

Knowledge of the few conventions and symbols below enable drawings to be more readily understood. They are based on the extensive recommendations in BS 1192:Part 3[3]. The application of some is illustrated in *fig 6.67*.

Levels and datums

Level on plans.

Level of cutting planes on sections and elevations.

Ceiling height above finished floor level (ffl) – shown on plans.

Bench Mark.

BM

Lines

Break line.

Line of section or cutting plane.

Beyond the cutting plane and visible.

Beyond the cutting plane but not visible.

Viewer's side of cutting plane and visible.

Viewer's side of cutting plane but not visible.

Direction of span of floor or roof structure.

Dimensions

Dimensions are normally shown in millimetres only*. This avoids confusion and the repetition of units, e.g.

10	not	10 mm
215	not	21.5cm or 0.215 m
1200	not	120 cm or 1.2 m
21000	not	21 m

*NOTE: When the United Kingdom changed to metric measurement it adopted the International System (SI) units which recommends the use of metres and millimetres. However, many other countries still use centimetres (cm) in addition.

Dimensions are normally written above and in the centre of a dimension line to be read from the bottom or right-hand edge of the drawing.

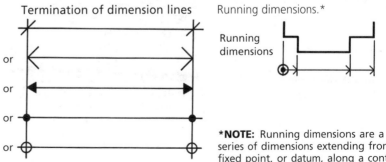

Termination of dimension lines

or

or

or

or

NOTE: Only one type of termination should be used on a set of drawings.

Running dimensions.*

Running dimensions

*NOTE: Running dimensions are a series of dimensions extending from a fixed point, or datum, along a continuous line on which each dimension is the distance from the datum.

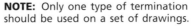

Steps and gradients

Direction of RISE of stairs, steps or ramp.

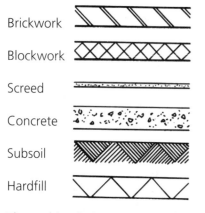

Straight stair/steps

NOTE: The numerals represent risers not treads. There is always one tread less than risers.

Symbols and hatching for materials

Brickwork

Blockwork

Screed

Concrete

Subsoil

Hardfill

Thermal insulation

Title and information panels

The following typical drawing title and information panel depicts the information normally included.

INFORMATION			
FOUNDATIONS:- 450mm WIDE TRENCH FILL.			
FOUNDATIONS:- MIN OF 1000mm BELOW GROUND LEVEL.			
EXTERNAL WALLS:- 102.5 FACING BRICK - 100mm CAVITY			
- 100mm BLOCKWORK.			
B 31.5.91 LIVING ROOM DIMENSIONS AMENDED.			
A 6.3.91 FLOOR LEVEL AMENDED.			
REVISIONS.			
ARCHITECT			
I.O.ZINE AND PARTNERS.			
JOB TITLE			
HOUSE FOR THE CLIENT.			
DRAWING TITLE			
GENERAL LAYOUT PLANS AND SECTIONS.			
JOB ARCHITECT			
M.I.JOBB.			
SCALE 1:100	DATE 25.2.91	DRAWN JK	CHECKED
JOB Nº 1001		DRAWING Nº 3	
		REVISION A B	

Abbreviations

Some of the customary and more common abbreviations used on drawings.

al	aluminium
BS	British Standard
c:l:s	cement:lime:sand
conc	concrete
dpc	damp-proof course
dpm	damp-proof membrane
ffl	finished floor level
galv	galvanised
grc	glass reinforced cement
hw	hard wood
max	maximum
min	minimum
mj	movement joint
rwp	rainwater pipe
s/s	stainless steel
sop	setting-out point
svp	soil & vent pipe
u/s	underside

KEY POINTS

- Select drawing required, consulting index if available.
- Ensure it is the latest amended version.
- Get an overall picture first before looking at details.
- Read information panels.
- Check that the sum of intermediate dimensions equals the overall dimension.
- Do not scale from drawings.
- Take every opportunity to compare drawings with the current construction.

USE AND CARE OF DRAWINGS ON SITE

One person on site should be responsible for the systematic receipt, recording and storage of drawings. In particular:

a. Record the number and date of each drawing and amendment issued.

b. Withdraw or mark superseded drawings. Always ensure work is carried out to the latest amendments.

c. Store where they will be protected and so that particular drawings can be readily retrieved.

d. Do not leave drawings in direct sunlight or they will fade.

In addition all users should:

- Set-out and build only from dimensions, **never** by scaling from a drawing. If vital dimensions are missing ask for instructions.

- As soon as possible check all dimensions relevant to the building operations for which you will be responsible and if there are any apparent errors or inconsistencies ask for instructions.

References

(1) BS 1192:Part 1:1984 'Construction drawing practice – Recommendations for general principles'.

(2) BS 1192:Part 2:1987 'Construction drawing practice – Recommendations for architectural and engineering drawings'.

(3) BS1192:Part 3:1987 'Construction drawing practice – Recommendations for symbols and other graphic conventions'.

(4) BS1192:Part 4:1984 'Construction drawing practice – Recommendations for landscape drawings'.

6.9   BRICKLAYING TOOLS AND EQUIPMENT

A bricklayer's varied kit of tools and equipment must be used correctly, safely, and must be well maintained, in order to produce high quality brickwork without wasting time and effort. This section describes their use and care.

BEDDING AND FINISHING TOOLS

Brick trowels

A variety of shapes, weights and lengths from 230 mm to 330 mm, are used for lifting and spreading mortar, removing the excess, finishing joints and rough cutting bricks *(fig 6.68)*.

- The largest may be preferred for building walls one-brick thick and more.

Figure 6.68. A variety of trowels.

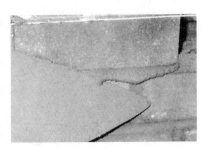

Figure 6.69. Removing excess mortar.

- Small trowels are recommended for intricate work.
- They are manufactured 'handed' left or right, although modern trowels often have both edges hardened for rough cutting.
- Choose one that has the right weight and balance for you. Do not assume that the biggest will be best.

Grip the trowel firmly with thumb on the ferrule, cut, roll, lay and spread the mortar to a nominal 10 mm thickness.

- Do not overload the bed joint. This most common mistake by trainees wastes time and mortar and stains the brick face.
- Position the brick, press down then run the slightly inclined trowel blade along the bed joint cutting the excess mortar cleanly and use for the next cross joint *(fig 6.69)*.
- When rough cutting never place your thumb on the side to be cut.

Clean in water daily or, in hot weather, at end of a shift.

- Remove build-up of mortar round the shank with a discarded piece of soft brick or emery cloth.
- Do not constantly tap bricks down on the bed with the end of the handle or it will become burred and uncomfortable to use. Metal caps on the end of the trowel

handles can damage brick faces.

Pointing trowels

These are obtainable from 175 mm down to 75 mm long. The latter is often termed a 'dotter' or dotting trowel and is used to fill and point cross joints.

Jointing tools

Special irons are obtainable for forming half-round tooled joints but a piece of hose of suitable diameter, a piece of wood or even a portion of a discarded bucket handle are used. But beware, a black rubber hose may cause staining of mortars. Keep the jointer flat *(fig 6.70)*. Using only the tip causes 'ribbing' marks *(fig 6.71)*. 'Tracking' is caused if the jointer does not cover the width of the

Figure 6.70. Using jointer correctly.

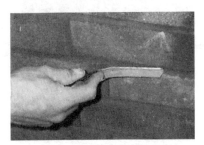

Figure 6.71. Using jointer incorrectly.

joint, either because it is too small for the joints or is not pressed in far enough.

Patent tools, having wheels, for moving freely over the brickwork, and interchangeable heads for forming recessed joints are known to bricklayers as 'chariots' *(fig 6.72)*. Use an appropriate width head to avoid tracking or damage to arrises.

Various patterns of 'chariot' jointing tools are available. The ones illustrated in *fig 6.72* and *fig 2.60* are typical.

Wash all jointing tools daily or more often in hot weather.

Brushes

Soft brushes are suitable for removing 'crumbs' of mortar after pointing or jointing. Use stiff brushes for removing mortar from recesses after raking or cutting out old mortar in repointing work.

Use dry brushes, as wet ones smear. Do not brush before the mortar is firm or use excessive pressure. Both can leave brush marks or smears. *(see fig 2.76 in Section 2.8 'Pointing & repointing')*

Wash brushes daily and allow to dry. Replace when worn.

TOOLS FOR CHECKING AND ALIGNING

Plumb rules

These have been superseded by spirit levels, but are still useful for checking the accuracy of spirit levels.

Spirit levels *(fig 6.73)*

1200 and 900 mm long levels of aluminium, plastic or hardwood are most commonly used, but 600 mm levels are used in

Figure 6.72. *A 'chariot' in use.*

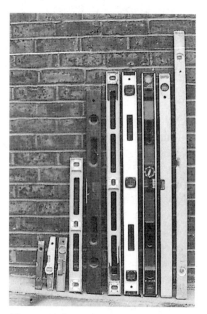

Figure 6.73. *Various spirit levels.*

restricted spaces and 250 mm boat levels for intricate work.

- **Never** use a hammer on a level to knock bricks into place.
- Clean daily, especially the bubble glass which tends to 'film over'.
- Check accuracy regularly by reversing the level on the same position or against a plumb level and adjust or replace bubble vials if inaccurate.
- Protect hardwood from wetting and swelling with boiled linseed oil.
- Take particular care of levels when working and travelling.

Lines and pins

Use hardened steel pins for securing building lines to brickwork joints. Cheap pins bend when being hammered in.

Building line is available in many materials.

- Polyester lines are brown or orange colour. They are cheap but fray easily.
- Hemp lines are traditional, can be spliced if they break but are prone to rot in damp tool bags. They are heavy and tend to sag over long distances.
- Cotton lines are similar to hemp but cannot be spliced.
- Twisted nylon lines are tough, tend to twist, fray and cannot be spliced.
- Woven nylon lines are tough, durable, do not sag greatly over long distances but stretch a lot.
- Most bricklayers prefer nylon lines.
- Put hitches into a line at regular intervals to prevent complete unwinding should the pin fall.
- After re-winding the line, tie a hitch on each pin to prevent its unravelling in the tool bag *(figs 6.74a,b & c)*.
- In wet weather open weave lines on pins so that they dry out quickly *(fig 6.75)*.

Corner blocks

Corner blocks can be quickly made from scraps of timber, or plastic ones may be bought *(figs 6.76 & 6.77)*.

- Thoroughly 'wind on' the line to the block as shown in *fig 6.77*.
- Keep blocks together in the tool bag with an elastic band.

Figure 6.74a. Lift.

Figure 6.74b. Twist.

Figure 6.74c. Tighten.

Figure 6.75. Open weave of line on pins.

Steel squares

Steel squares are particularly useful for checking the squareness of piers. They should be wiped clean daily.

Tapes, rules, pencils

- Steel tapes are compact and long but get full of grit.

- Steel tapes should be wiped clean daily and refilled when worn or damaged.
- Folding rules are shorter and more rigid but get trodden on. It is a matter of personal choice.
- A pencil is essential for marking perpends but take care in use as pencil marks on some facing bricks are difficult to remove. A 2H pencil lasts longer than an HB. *(see Section 2.1 'Setting-out facework')*

CUTTING TOOLS

Club hammers *(fig 6.78)*
Sometimes known as lump hammers, are available in 1 kg and 2 kg weights for use with cold chisels and bolsters. The 1 kg hammer is better for most work.

Do not hit bricks and blocks with the end of the handle nor the faced sides with the hammer head. The former will burr the handle and be uncomfortable, the latter will damage the face of bricks and blocks.

- Wipe hammers dry daily.
- Check head regularly and if loose remove wedges, re-wedge handle or fix a new one. As a temporary measure only, soak in water to tighten head.

AN AWFUL WARNING

Never hit two hammer heads together to hear the ring or feel the bounce. Sharp pieces of metal can fly off at great speed, severely slicing any body they hit.

Figure 6.76. Wooden corner block.

Figure 6.77. Plastic corner block.

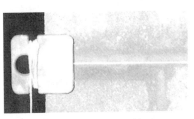

Figure 6.78. Club, brick hammers and scutches.

Figure 6.79. Bolsters and club hammer.

Cold chisels and bolsters *(fig 6.79)*
Plain chisels 15 to 28 mm wide by 250 mm long and bolsters 50, 75 and 100 mm wide are available.

- The 100 mm bolster is best for cutting bricks.
- Comb chisels, similar to a scutch, are useful for chasing or trimming existing walling.
- Plastic mushroom sleeves may be fitted to most chisels to reduce the risk of injury to the hand.

Wear eye protectors, hold club hammer firmly at end of handle, not halfway. Position the chisel and, for maximum accuracy, swing the forearm only, not the whole arm. When cutting bricks do not hit too hard, firm light taps are sufficient and cause less damage if you miss *(fig 6.80)*.

All cold chisels should be kept sharp, preferably in a metal workshop where they can be tempered. The opportunity should be taken to remove burred or mushroomed heads with a grindstone.

Brick hammers

A brick hammer has a chisel blade at one end for cutting and a square end for knocking in nails *(fig 6.78 second from left)*.

***Figure 6.80.** Cutting a brick.*

It is particularly useful for rough cutting hard bricks where a trowel would be too light and might be damaged.

Take care as for a club hammer, but in addition have the chisel end sharpened and tempered regularly.

Scutches *(fig 6.78)*

A scutch may have two grooved ends to receive replacement steel blades or replacement combs for trimming or cutting. Alternatively they may have one end grooved and the other square like a brick hammer. The blades are simply knocked out sideways and turned round or replaced as necessary.

Care for them as for a brick hammer, but no sharpening is required.

Keeping tools together

Consider storage arrangements for care and security, including locking of vehicles, a tool lock up, and a canvas bag for the day's needs. The disadvantage with a bucket for this purpose is that during a heavy shower you collect water.

ADDITIONAL EQUIPMENT FOR BRICKLAYERS

Patent metal corner profiles *(fig 6.81)*

These can be obtained in sets of two or more, marked with the standard vertical gauge.

They are quickly bolted in position with the lines held by simple sliding clips which can be raised as each course is built.

- No quoins, as described in section 2.3, are necessary.

***Figure 6.81.** A corner profile setup.*

- Profiles are particularly effective when building with irregular bricks such as hand made and soft mud bricks with which the building of corners is difficult.
- All bricks are laid to the line, and a spirit level is required only to check that the profiles are plumb, which should be done regularly.
- If the permanent gauge markings are unsuitable for a particular job, mark out the required gauge with masking tape and remove at end of job.

Lock away at night for safety.

- Do not mishandle or leave lying about as they become useless if bent.
- Clean daily, especially any mortar around the screw threads. Oil the latter weekly.

Patent profiles of various effective designs are produced by different manufacturers. The tool illustrated in diagrams in this book is only intended as a typical example and no inference should be drawn that this particular design is approved or preferred by the authors.

Figure 6.82. *Using hand saw.*

Some other designs have additional features and/or accessory fittings that extend their usefulness in assisting accurate and well controlled bricklaying.

Tungsten tipped hand saws

These are used for cutting out bricks from existing walling to form toothings, providing that hard, cement-rich mortars have not been used. Hand saws are often preferable as they are vibration free and quiet but take longer than mechanical saws *(fig 6.82)*. They are useful for cutting lightweight blockwork particularly up a gable as they require no power or trailing leads.

Avoid hard materials which can knock out the teeth.

- Start gently with small cuts to a depth of 10 mm as they have large teeth and are liable to jump.
- When nearly through, slow down and catch the off-cut, especially if working on scaffolding.

Electric saws with tipped reciprocating blades

Developed from wood cutting saws to cut concrete blocks quicker, they require a 110 volt power supply and hence have trailing cables.

Wear goggles against flying particles. It is a dust free operation and does not require a mask.

Small angle grinders

This normally has a 100 mm diameter masonry cutting blade and is commonly used to remove existing mortar before repointing.

- Wear goggles and mask against dust.
- Wear gloves against abrasive grit particles.
- Wear ear protectors especially in confined spaces.
- Use correct blade i.e. silicon carbide type abrasive **not** aluminium oxide.
- It is most important to ensure that the speed on the blue label matches the maximum speed of the angle grinder.
- Unplug the machine before fitting a new blade.
- Fit the blade correctly and secure tightly.
- Ensure that the electrical supply is not a hazard to the operator or others.
- Do not use above shoulder height. Keep away from body, preferably at arm's length.
- Position guard to deflect particles from face.
- Make sure that the wheels on all portable machines are reinforced to prevent breaking up under centrifugal force.
- When cutting small units like bricks and tiles provide a stop to restrain the brick when the blade is applied. A batten nailed to a board will suffice.
- Do not start with blade in contact with unit, it may

snatch and throw the unit or you off balance.

- Hold the unit firmly, preferably with a cramp.
- Operate in a place free of passing people and debris underfoot.
- Never put an angle grinder down until the blade stops spinning.

NOTE: A safer and more accurate way of cutting small units is on a masonry bench saw.

Masonry bench saws *(fig 6.83)*

They may be electric or petrol driven and consist of:

- A powered cutter moved by hand, treadle or both.
- A moveable trolley on rails with a stop and sometimes an adjustable stop for angle cuts.
- A water supply pumped to the blade for dust control. *(see Section 2.5 'Cutting bricks' for operating information)*

Figure 6.83. *(see also Figure 2.48) Bench saw.*

Larger portable angle grinders *(fig 6.84)*

Capable of taking larger blades from 200 to 300 mm.
Used for:
- Cutting masonry units.
- Chasing walls for services.

Figure 6.84. *Angle grinder.*

- Cutting holes for openings or for toothing block walls.
- The deeper the cut the larger the blade required.
- Machines for large blades are heavier and less manoeuvrable.
- **A person changing the blade on a machine capable of taking a blade exceeding 235 mm must have attended an abrasive wheels training course.**
- Beware of 'snatch' when first starting large machines.
- Where a neat straight cut is required fix a guide to the wall. Set the batten so that the guard, not the blade, runs against it to avoid cutting the batten.

SAFETY

Building sites are dangerous places and bricklayers' bodies as liable to serious injury as anyone's. It is common sense to be constantly aware of this and take precautions.

Eyes

Keep goggles hung round the neck ready for use when cutting. When using mechanical plant a helmet and visor are advisable.

Heads and enclosed brains

Since 1st April 1990 everyone on construction sites has been required by law to wear helmets. Adjust the back strap to keep them on when bending forward. Alternatively fix a chin strap.

Feet, toes and ankles

Wear safety boots with steel toe caps to support ankles and prevent broken toes.

Hearing

Wear ear protectors when using noisy tools and plant.

Lungs

Wear respirators or masks where dust is generated, especially in confined places.

Skin and flesh – infection and abrasion

Regular contact with cement or lime can cause allergic reactions. Cuts from sharp and abrasive materials can infect the flesh. Get immediate First Aid attention to clean and cover cuts. Wear barrier cream and/or gloves.

First-aid kits

Even though these are kept on sites it makes sense for bricklayers to carry their own small kit or at least some adhesive plasters.

KEY POINTS

- Use only the correct tool for the job. Ensure that it is in a safe condition.
- Use all tools correctly to produce quality work efficiently and safely.
- Clean and dry tools daily, sharpen as necessary.
- Replace all tools if they become inefficient or unsafe.

6.10 BRICK MANUFACTURE

Bricks have been made in many parts of the world for thousands of years. At first, clay was moulded and dried in the sun. Eventually, brick makers learned to make harder more durable bricks by firing. In recent times, sand with lime and rock aggregates with cement have been used (fig 6.85).

During the last 25 years brick making has changed from being predominantly manual to being highly mechanised using modern technology. In this way bricks remain competitive in terms of appearance, performance, productivity and fuel efficiency.

This section aims to provide an understanding of how the physical and chemical properties of bricks, like

Figure 6.85. *Relative numbers of bricks made in the UK.*

91% Clay

7% Concrete 2% Calcium silicate

strength and durability as well as appearance, depend mainly on the type of clay or other material and the method of manufacture used. But it is not possible, simply by reference to the type of materials and manufacturing methods used, to predict a bricks characteristics. Instead it is necessary to understand the very precise terminology used to describe them. Nor is it practicable for manufacturers, using a given brick material and method of manufacture, to change the physical properties from those established by test and declared in their catalogues.

BRICK STANDARDS, CLASSIFICATION AND QUALITY

British Standard Specifications
In the UK bricks are specified in terms of physical properties such as durability, water absorption, compressive strength, form and dimensions, in three British Standard Specifications: BS 3921[1] for clay bricks; BS 6073[2] for concrete bricks; BS 187[3] for calcium silicate bricks.

Within the next few years these British Standards will be replaced by European Standards.

Classification
The following physical properties are determined by tests on samples of bricks as described in the appropriate British Standard Specifications listed above.

Compressive strength
Measures the resistance to crushing of bricks, expressed in Newtons per square millimetre (N/mm^2) and may range from 10 to well over 200 N/mm^2 for different types of bricks. The information is used by engineers to calculate the strength of structural brickwork. It is also a property used to classify classes of clay engineering bricks and classes of calcium silicate bricks *(see Section 6.4 'Durability of brickwork')*. Compressive strength is **not** in general an indication of the frost resistance of a clay brick.

Water absorption
Measures the increase in weight, expressed as a percentage, of a brick when saturated, compared with the same brick when completely dry. The water absorption of the most dense to the least dense bricks ranges from less than 4.5% to over 30%. Water absorption is a property used to classify clay engineering and DPC bricks *(see Section 6.4 'Durability of brickwork')*. Water absorption is **not** in general an indication of resistance to frost attack or resistance to rain penetration. *(see Section 6.7 'Rain resistance of cavity walls')*

Resistance to frost attack
Frost resistance is determined by test in the laboratory and from experience in use. Clay bricks are classified as Frost Resistant (F) or Moderately Frost Resistant (M). *(see Sections 6.2 'Frost attack and frost resistance' and 6.4 'Durability of brickwork'.)* The frost resistance of clay bricks is **not** in general related to either compressive strength or absorption.

Soluble salts content
The soluble salts content of clay bricks measured in accordance with BS 3921[1] is expressed either as Low (L) or Normal (N). *(see Section 6.4 'Durability of brickwork')*

Clay engineering bricks
Have high strength and low water absorption and are classified as Class A and Class B. They are not necessarily manufactured to the standard of facing bricks *(see Section 6.4 'Durability of brickwork')* **Note**: there is no standard classification for the so-called semi-engineering brick.

Clay damp-proof course (DPC) bricks
Are low absorption bricks suitable for use as DPCs. *(see Sections 6.4 'Durability of brickwork' and 4.3 'Damp-proof courses')*

Facing units
Are manufactured to give an attractive and consistent appearance.

Common units
Are suitable for general construction work where their appearance is unimportant.

Bricks rejected from other grades may be classified as such. (The last two classifications are defined in BS 6100[4]).

Recommendations for use of bricks of particular properties for particular applications are made in the Code of Practice for masonry BS 5628:Part 3[5].

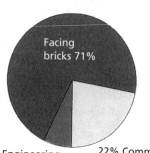

Figure 6.86. *Relative numbers of bricks currently produced in the UK.*

Standard and special shapes and sizes

The vast majority of bricks are manufactured to the standard 'work sizes' of 215 × 102.5 × 65 mm. *(see Section 2.1 'Setting-out facework')*

. Bricks of rectilinear or cuboid form, other than the standard sized one defined in BS 3921 and BS 147, are listed in BS 4729 'Dimensions of bricks of special shapes and sizes'. This standard also defines the shapes and sizes of non-rectilinear bricks that are used to form angles, curves, features and other construction details which cannot be achieved satisfactorily with standard bricks. *(see Section 2.9 'Bricks of special shapes and sizes')*

VARIATIONS IN BRICK SIZES

Even with the best quality control there remain slight variations in raw materials and manufacturing processes – after being formed bricks shrink during drying, firing, autoclaving and curing. Consequently, the 'actual' sizes of bricks made to a British Standard are permitted to vary from the 'work sizes' within limits specified in the standards. *(see Section 6.5 'Allowing for variations in brick sizes')*

Quality control

During the 1980s the majority of brickworks achieved and have maintained accreditation to BS 5750[7] (ISO 9000/EN 29000) 'Quality Management Systems'. This means that the factory management system has been independently assessed to conform to the standard which covers every aspect of the manufacturing process from raw materials to delivery of the finished bricks.

CLAY BRICKS

Raw materials

Brick clays are relatively soft sedimentary rocks known by names such as clays, shales, mudstones and marls. Clays are plastic, i.e. they are readily moulded to a given shape which will be retained. But adequate plasticity of excavated clay is frequently only developed by crushing, grinding and mixing with water.

Brick clays consist largely of quartz and clay minerals. The latter, being amongst the finest particles, are responsible for the plasticity.

Brick clays are converted into hard durable bricks by firing. At the highest temperatures of 900°C–1250°C in the firing process, partial melting or vitrification of the components of the clay minerals occurs. On cooling a glass develops which binds the material together.

Because brick clays shrink significantly on both drying and firing the process is controlled to minimise cracking and variations in the sizes of finished bricks.

Inert minerals, like quartz, help to prevent excessive shrinkage during drying and excessive melting and shrinkage during firing. Other minor constituents of the clay mineral, e.g. sodium and potassium, assist vitrification.

Iron compounds, e.g. iron oxide as haematite, are responsible for the predominant red fired colours.

Many other minerals like carbonates, e.g. limestone, chalk, dolomite; soluble salts, e.g. calcium sulfate as gypsum; and carbonaceous materials, e.g. coal, lignite, may well affect the characteristics of the finished bricks and manufacturers have to control them.

Table 6.6 below lists the key brick manufacturing clays used in the UK and their locations.

Most clays are sufficiently soft to be excavated mechanically without blasting. Selection of the most appropriate equipment will depend on the characteristics of the deposit and the method of working the quarry face.

Many manufacturers build stockpiles to avoid the difficulties of moving vehicles in wet clay during winter.

Working methods depend on the characteristics of each

TABLE 6.6

	Age (millions of years)	Environment of deposition	Areas used for brickmaking
Boulder clay	<1	Glaciers & rivers	N England
Brickearth	<1	Land	S E England
Reading beds	circa 65	Deltas	S England
Gault clay	circa 90	Marine	S E and East Anglia
Weald clay	circa 130	Deltas/lakes	S E England
Oxford clay	circa 160	Marine	Central/East Anglia
Keuper marl	circa 180	Salt lakes	Midlands
Etruria marl	circa 270	Fresh water	Midlands
Coal measure shales	circa 310	Deltas and swamps	S Wales, Midlands, North Scotland
Coal measure fire-clays	circa 310	As above	As above
Culm measures	circa 310	Marine	S W England
Devonian shale	circa 390	Marine	S W England

deposit and the need for the clay fed to the brick plant to be consistent and predictable.

Clays, being sedimentary, occur in layers. Where materials of different characteristics are being combined in the same clay mix, it is common practice to build a layered stockpile (fig 6.87) which may contain sufficient raw material for a period of 12 months. When clay is removed for delivery to the brick plant a full vertical cut is taken from the face.

Clay preparation

The aim of clay preparation is to deliver to the shaping machinery a body of clay consistent in content, grading, plasticity and water content. If the clay feed varies, so will the finished bricks.

A sequence of machines grind and work the clay to obtain plasticity and uniform workability to suit the shaping equipment. Coarse clay is crushed, ground, rolled, cut, kneaded, etc. with appropriate additions of water ready for the forming machine.

Primary crushing reduces large pieces to 75–100 mm.

The dry pan mill (fig 6.88) is widely used for secondary grinding between heavy large diameter rollers and plates. The largest particles are now typically 3–5 mm in diameter.

The grinding process yields a range of particle sizes to achieve the required packing density and porosity. The fineness of grinding influences not only the internal and external textures of bricks but also characteristics such as strength, durability and water absorption. The effect of particle size on appearance is particularly evident in dragfaced products.

Mixing

After grinding, the moisture level is increased for forming. The clay body is worked to make it homogeneous, with the water evenly distributed through the clay particles.

The double shafted mixer is the most widely used for plastic clays. Two horizontal shafts fitted with overlapping knives rotate in opposite directions. The clay is cut and kneaded as it is driven from one end of the mixer to the other.

Additions

Materials to produce through colours (e.g. manganese dioxide to produce brown or grey bricks)

Figure 6.87. A layered stockpile.

Figure 6.88. A dry pan mill.

are usually added at the mixing stage.

More substantial additions of, for example, sand or coke are sometimes made to control shrinkage, act as a fuel or generate a specific appearance. These are generally incorporated into the clay mix via feeders prior to the mixing stage.

Shaping

The forming processes used to manufacture the majority of bricks in the UK fall into three categories: soft mud moulding by hand or machine; extrusion/wire-cut; semi-dry pressing.

The clay bodies used for each process are characterised by fundamentally different moisture contents and workabilities.

Shrinkage occurs from the wet (formed) through to fired stages of brick production. Bricks are, therefore, initially formed larger than the intended work size.

Soft mud moulding to produce stock bricks

THE TRADITIONAL HANDMAKING PROCESS
The handmaker forms a roughly shaped clot from a mix of a soft mud consistency (greater than 20% moisture). The clot is coated with sand and thrown into a mould, generally precoated with sand (fig 6.89). The bottom of the mould is formed by the stock. A kicker may be placed on the stock to form the frog. The stock gives its name to soft mud sand moulded bricks whether hand or machine made. The thrown soft plastic clot adopts the general shape of the mould and the crease patterns characteristic of these products and known as 'the handmaker's walk'.

The sand coating of the mould allows release of the brick from the mould for drying. Soft mud bricks are too soft to stack after moulding and are dried individually on pallets.

Frogs assist forming, drying and firing.

MACHINE MOULDING
Machines for moulding stock bricks have outputs as high as 22,000 bricks per hour. A practised handmaker may typically make around 100 bricks per hour.

In the standard process, a clay mix of soft mud consistency is forced through steel dies by spiral shaped blades into sanded moulds (fig 6.90). Brick faces are smooth rather than creased.

A development of the standard process machines throws sanded clots of clay simultaneously into a series of moulds. The configuration of each clot is different and the sanding varies continuously so that no two bricks are identical. This close simulation of hand throwing creates crease patterns on the brick faces.

'Slop moulded' or waterstruck' products are also manufactured from a soft mud mix. Release from the mould is facilitated by water rather than sand giving a different and characteristic surface texture.

Extrusion/Wire-cut

In the UK this method accounts for around 40% of production.

A plastic clay mix is driven through a die with a screw. A continuous column is formed with a cross section based on the 215 × 102.5 mm dimensions (fig 6.91). The moisture content of

Figure 6.89. *Throwing a brick by hand.*

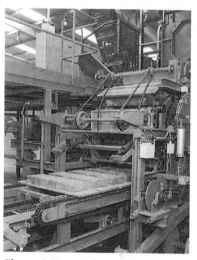

Figure 6.90. *A soft mud brick machine with an output of 6,000 bricks per hour.*

Figure 6.91. *An extruder and column.*

the clay body is generally in the range of 15–20%. Outputs of 20,000 bricks per hour can be achieved through a single die.

The perforations in extruded bricks are formed by core bars in the die head. The size and number of perforations vary from works to works but, in the UK, do not exceed 25% of the total volume of the brick. Perforations reduce drying and firing costs and incidentally reduce the weight.

Surface texturing to generate rustic, brushed, dragwire, rolled or sanded finishes is carried out immediately after extrusion. Colours may also be added at this stage by stains applied with the sand or via a spray.

Next, the column is cut into a slug which is cut into bricks via a multiple wire cutter *(fig 6.92)*.

Semi-dry pressing

The semi-dry pressing process is used to form Fletton bricks, accounting for a significant proportion of the UK brick output. The term 'semi-dry' refers to the workability of the clay which has a natural moisture content of around 17–20%. A fine granular material is produced by grinding the raw clay and pressing the

ground material into bricks without reducing the natural moisture content.

Compaction is by direct application of pressure in a mechanical press *(fig 6.93)*. This helps to develop maximum consolidation and compaction of the clay. Fletton bricks have frogs pressed into them to assist the drying and firing of the bricks.

Semi-dry pressing requires the minimum amount of moisture for any particular clay and so drying costs are reduced.

Other forming methods

Stiff plastic pressing involves a combination of extrusion to form a roughly brick-shaped clot and then pressing in a mould.

Extruded wire-cut bricks may be repressed to obtain a particular surface finish or chamfer, etc.

Handling and setting

After being formed, bricks are dried to specific moisture levels and set in appropriate patterns on kiln cars or static hearths ready for firing.

The bricks are moved and handled increasingly by mechanisation in order to maximise productivity *(fig 6.94)*.

Some bricks, like stiff extruded wire-cut bricks and Flettons, can be set directly into firing patterns onto a kiln car, by hand or by setting machines, for drying and firing without an intermediate handling stage.

Drying

The drying of clay bricks is technologically complex and important for two reasons:

- The shrinkage which takes place (5–14% depending on the clay and moisture

Figure 6.92. *A typical multiple wire cutter.*

Figure 6.93. *A semi-dry brick press.*

Figure 6.94. *Extruded bricks being marshalled onto pallets ready for drying.*

content) can cause cracking if the process is not effectively controlled.

- The energy used may be as much as 30–50% of the total requirement. Efficient use of energy is essential to minimise costs though much of the heat may be in the form of hot air transferred from the kiln to the dryer.

Firing

Firing imparts colour, strength and durability to bricks. As it is a significant cost in terms of fuel consumed, energy efficiency is a high priority. Firing is carried out in a range of types and sizes of kilns, depending on the output required and the type of product.

Kilns are fired by burning carbonaceous fuels such as gas, coal, coke and oil.

Bricks undergo a sequence of fundamental changes as the temperature builds up during the firing process. The most important of these are:

- **Up to 150°C**
  Any residual water from the drying process is removed.
- **150°C–650°C**
  Clay minerals break down to give off water.
- **200°C–900°C**
  Burn out of carbonaceous material which may be present in the clay or added as, for example, coke. Carbon can be important both in its contribution to the appearance of a brick and as a fuel.
  However, carbon remaining at the highest temperature can contribute to bloating and it is vital to

control the rate of temperature rise.
- **900°C–1250°C**
  At the highest temperatures liquids form as a result of partial melting of components of the clay minerals. On cooling the liquids form a glass which binds the brick into a hard and durable unit. The characteristic colours develop at this stage. In a normal kiln atmosphere the majority of clays will fire to a red colour but fireclays, for example, yield buff/cream coloured bricks. These basic colours can be modified by adjusting the fuel:air ratio or by ensuring that some carbon is retained in the brick body up to the highest temperatures.
- **Cooling**
  At 573°C silica, a major component of the body, undergoes a change in volume. The rate of temperature change through this zone is controlled, especially on cooling, otherwise internal cracking can occur.
- **Shrinkage**
  In addition to the shrinkage which takes place on drying, further shrinkage occurs during firing.

Kilns

Kilns can be divided broadly into two categories:
- CONTINUOUS/SEMI-CONTINUOUS Always in some part of the kiln bricks are being fired, unfired bricks are being introduced and fired bricks are being withdrawn. Continuous kilns are more fuel-efficient than

intermittent kilns. Over 90% of UK clay bricks are fired in this type of kiln.
- INTERMITTENT – the kiln is loaded with bricks which are then fired, cooled, and removed. The cycle is then repeated.
  Intermittent kilns are used to achieve specific characteristics in the products or to fire small quantities of bricks, for example, special shapes.

INTERMITTENT KILNS
a) CLAMPS
The clamp method has a long history. It is still used today, to a limited extent, for firing stock bricks in S E England.

Essentially, a clamp is a large stack of closely set bricks containing fuel. It is ignited at one end and left to burn. The fire gradually proceeds along the full length of the clamp.

There may be over 1 million bricks in a clamp. It is hand built and situated outdoors, or with a simple roof cover. The whole stack is laid on a bed of fuel (usually coke) supported, typically, by a couple of courses of bricks already fired. There is sufficient fuel in the bricks (coke, old refuse, etc.) to raise the temperature to over 1100°C. Some modern clamps are initially fired by gas instead of a bed of coke.

Firing is a lengthy process and the whole cycle may take over 6 weeks. On completion the bricks are withdrawn and sorted/packed on jigs raised into appropriate positions.

Clamp firing produces a wide range of colours for a relatively low initial capital cost.

Figure 6.95. *A burning clamp of bricks.*

Once a clamp is ignited not a great deal can be done to control it. Some parts reach higher temperatures than others and sensitivity to climatic conditions adds further variables. Yields of first quality 'best' bricks can be relatively low.

b) MOVING HOOD KILNS
The moving hood kiln is a recent development. It is finding increasing use on small/medium volume stock brick plants as well as for firing special bricks of all types.

The kiln runs via motorised rollers on rails between two fixed hearths. At the completion of a firing the hood will move over the second hearth already loaded with green, dried bricks *(fig 6.96)*. Firing will commence immediately

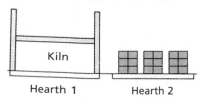

Figure 6.96. *Principle of moving hood kiln operation.*

and be completed in around 3 days. The fired bricks from the first hearth are unloaded and the hearth filled with green dried bricks to await firing.

When used for stock bricks, energy for firing the bricks comes both from the gas fired burners in the kiln and the coke present in the brick body. Each individual dense set stack of bricks behaves in a similar manner to a clamp and is capable of yielding a similar rich spectrum of colour.

c) SHUTTLE KILNS
Special shaped bricks are more difficult to fire than standard bricks. It is often not possible to fire different sizes and shapes together and special support structures are sometimes necessary to ensure stability during drying and firing. Special bricks and small quantities of standard bricks are frequently fired in gas fired intermittent shuttle kilns. Products are set on kiln cars which are pushed into the kilns, fired, cooled and withdrawn for sorting and packing.

Control
The key changes which take place as the temperature rises during the firing of clay bricks have been listed earlier. In conjunction with several other factors such as the mineralogy of the clay mix and the permeability of the body, these determine how rapidly a brick can be heated through each temperature zone. The faster the bricks can safely be heated the lower the fuel consumption.

A firing profile is compiled for each body mix on a particular plant *(fig 6.97)*.

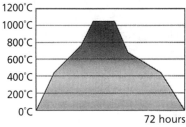
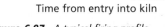

Figure 6.97. *A typical firing profile.*

The objective is to control a firing in line with the target profile. Temperatures are determined by thermocouples situated at strategic positions. When action is required in response to temperature data from thermocouples it may be effected:

1. Manually
2. Via a network of individual electronic controllers
3. Via a network of individual electronic controllers managed by a computer *(fig 6.98)*.

CONTINUOUS KILNS
There are two main types of continuous kiln: a) tunnel kilns and b) chamber kilns.

Figure 6.98. *Kiln control panel on a modern brick plant with computer managed tunnel kiln firing.*

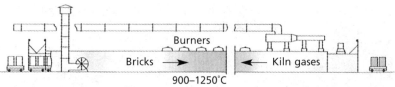

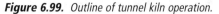

900–1250°C

Bricks to be fired are set on a deck of refractory bricks. This is then seated on an iron/steel base with two sets of wheels

The car wheels run on rails. This part of the kiln is protected from the hot kiln gases by the deck refractories and seals on the sides

Figure 6.99. *Outline of tunnel kiln operation.*

Figure 6.100. *Gas fired product emerging from a typical modern tunnel kiln.*

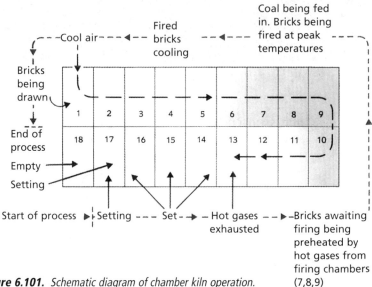

Figure 6.101. *Schematic diagram of chamber kiln operation.*

a) TUNNEL KILNS

- Cars loaded with bricks move through the fire. Around 55% of UK brick production is now fired in tunnel kilns. The fuel used is, almost without exception, gas.
- Tunnel kilns first became commercially successful around 1910. The basic concept is illustrated in *fig 6.99*.
- At intervals of, for example, 75 minutes, a car of unfired bricks is pushed into the kiln and a car of fired bricks emerges *(fig 6.100)*. The cars are pushed intermittently into fixed positions and burners in the roof and/or on the sides, fire into the gaps or dykes between the cars. Typically there may be 40 cars in a kiln

and the total time for each one to pass through the kiln may be around 2½ days.

- The peak temperature and rate of heating and cooling may be adjusted to suit any particular brick range.

b) CHAMBER KILNS

- The fire moves progressively round the kiln. About 35% of UK brick production is fired in chamber kilns. The majority of this volume is of semi-dry pressed Flettons and coal is the predominant fuel.
- The prototypes of modern moving fire, or annular, kilns were developed in the mid-

nineteenth century. The basic principle of operation is illustrated in *Figure 6.101*.

- The majority of kilns of this type are fed by coal through feed holes in the roof of each chamber. Much of the fuel for firing Fletton bricks comes from carbonaceous material which occurs naturally in the clay.
- The operation is a continuous circuit of setting, drawing and moving forward of the firing zone from chamber to chamber.
- When the fired bricks are cool enough the next wicket will be taken down, stacks of

Figure 6.102. *Unloading stacks of fired bricks from a chamber kiln. The open wicket is behind the forklift truck.*

Figure 6.103. *Fully mechanised unloading of fired stock bricks from a kiln car.*

Figure 6.104. *Fired bricks are mechanically offloaded from a kiln car onto an inspection belt where defective bricks are removed.*

fired bricks will be withdrawn *(fig 6.102)* and, in due course, stacks of green bricks will be set.

Handling and packing of fired product

The key objectives at this stage of the process are:

- To sort product into different grades and remove defective bricks.
- Where necessary to blend bricks, especially multis to ensure minimum variation from pack to pack.
- To minimise costs. Where packing is manual this can represent a significant component of the total labour costs of a factory.

Where unloading from kiln cars and packing are fully mechanised *(fig 6.103)* systems are still installed to allow all bricks to be individually inspected and defective bricks to be removed.

Packs of bricks are often supplied wrapped in polyethylene sheet to give protection from the weather during transport and storage on site.

CONCRETE BRICKS

- Essentially concrete bricks are composed of aggregates (e.g. crushed limestone, granite, etc.) bonded with cement and coloured with pigments.
- The bond is formed as a result of a chemical reaction between the cement and water. The strength of this bond increases steadily with time and continues after the product is built into place.
- The mix of aggregates, cement, water (and pigment[s]/admixture[s] as

required) is compacted into bricks by pressure, vibration or high frequency hammer action. In the UK the most widely used method is application of pressure using mechanical or hydraulic presses.

- The pigments used are predominantly iron oxides which are totally durable and colourfast.
- Brick presses operate continuously but mixing is carried out on a batch by batch basis.

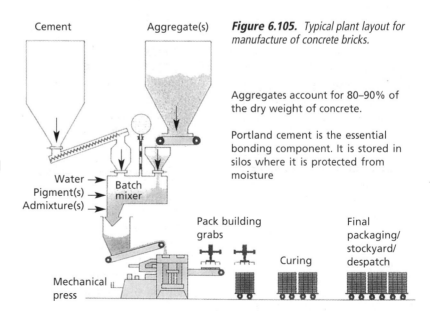

Figure 6.105. *Typical plant layout for manufacture of concrete bricks.*

Aggregates account for 80–90% of the dry weight of concrete.

Portland cement is the essential bonding component. It is stored in silos where it is protected from moisture

Cement

Aggregate(s)

Water →
Pigment(s) →
Admixture(s) →
Batch mixer

Mechanical press

Pack building grabs

Curing

Final packaging/ stockyard/ despatch

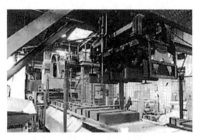

Figure 6.106. *Mechanical press for concrete bricks.*

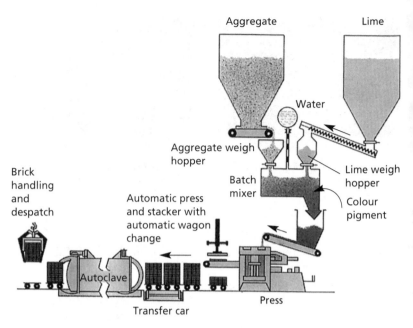

Figure 6.107. *Typical plant layout for manufacture of calcium silicate bricks.*

- Mechanical/hydraulic presses are fed with semi-dry concrete. The moisture content is in the range 4–7% and the water addition is carefully controlled to ensure a product of consistent strength and water absorption.
- The mechanical press in *fig 6.106* produces 8 bricks per cycle (about 4000 bricks per hour). A rustic finish is created by brushing the stretcher faces whilst the bricks are still in the mould. Grabs build green bricks directly into the finished pack configuration.
- Some concrete bricks have frogs or perforations. But the majority of facing bricks are solid and heavier than clay facings.
- Smaller numbers of concrete bricks are produced on machines which compact primarily by vibration.
- Size changes from pressed to cured bricks are negligible. Dimensional consistency is therefore a characteristic of concrete bricks.
- When pressed, concrete bricks are weak but strength increases steadily with time. This process is referred to as curing. Bricks are only

released for sale when they have reached the minimum strengths specified.
- Concrete facing bricks are generally shrink-wrapped and the products also need to be covered on site.

CALCIUM SILICATE BRICKS
- Calcium silicate bricks are also referred to as 'sandlime' or 'flintlime' bricks.
- Calcium silicate bricks are composed of aggregates with a hydrated calcium silicate bond and colour derived from pigment additions.
- The mix is moulded under high pressure in a mechanical or a hydraulic press.
- The bond is formed as a result of a chemical reaction between lime and a siliceous aggregate. The reaction is promoted by curing in autoclaves using steam at

elevated pressures and temperatures. This yields a strong, durable product.
- A typical plant layout is shown in *fig 6.107*.
- The aggregates account for about 90% of the dry weight of the mix.
- Hydrated lime is stored in silos in powder form or generated on site by hydration of quicklime in a reactor.
- The pigments used are predominantly iron oxides which are totally durable and colourfast.
- Brick presses operate continuously but mixing is carried out on a batch by batch basis.
- The majority of calcium silicate bricks are frogged and weights are comparable to clay bricks.
- Pressed 'green' bricks are loaded onto steel trolleys and subjected to high-pressure

Figure 6.108. Calcium silicate bricks emerging from an autoclave and ready for despatch.

steam for 5–12 hours in autoclaves *(fig 6.108)*. After packaging the bricks are ready for immediate despatch.

- Size changes from pressed to cured bricks are negligible. Dimensional consistency is therefore a characteristic of calcium silicate bricks.

- Property requirements and methods of classification for calcium silicate bricks are specified in BS 187:1978. Five strength classes are listed from 20.5 N/mm^2 to 48.5 N/mm^2. Calcium silicate bricks which meet this standard are frost resistant.

- Calcium silicate bricks have negligible soluble salt contents and are not, therefore, prone to efflorescence.

References
(1) BS 3921:1985 British Standard Specification for *'Clay Bricks'*.
(2) BS 6073:Part 1:1981 *'Precast Concrete Masonry Units'*.

Specification for precast concrete masonry units.
BS 6073:Part 2:1981 *'Precast Concrete Masonry Units'*. Method for specifying precast concrete masonry units.
(3) BS 187:1978 *'Specification for calcium silicate (sandlime & flintlime) bricks'*.
(4) BS 6100:Section 5.3:1984 *'British Standard Glossary of building and civil engineering terms'*. Part 5. Section 5.3 *'Bricks and blocks'*.
(5) BS 5628:Part 3:1985 British Standard Code of Practice for *'Use of Masonry'*. Part 3. *'Materials and components, design and workmanship'*.
(6) BS 4729:1990 Specification for *'Dimensions of bricks of special shapes and sizes.'*
(7) BS 5750:Part 2. Quality Systems. *'Specifications for manufacture and installation.'*

6.11 BLOCKWORK INNER LEAVES, WALLS AND PARTITIONS

This section deals with blockwork inner leaves and internal walls and partitions. Building with facing-quality blocks and common quality blocks, fair-faced, is beyond the scope of this section.

Types of blocks
Concrete blocks are made and specified to the requirements of BS 6073[1][2] in three basic forms *(fig 6.109)*. **Solid** blocks have no formed voids, **cellular** blocks have one or more which do **not** pass right through the block and **hollow** blocks have one or more which do pass through the block.

Some manufacturers make blocks with an insulant bonded to the outside of one face or inserted in the voids.

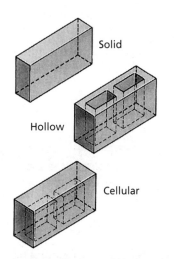

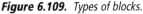

Figure 6.109. Types of blocks.

Typically, the nominal densities of blocks range between 475 and 2200 kg/m^3. The least dense, and usually lightest, being autoclaved aerated blocks and the most dense, and usually heaviest, being solid dense natural aggregate blocks. There is a wide range of densities and weights between these extremes.

Special shapes and sizes include cavity closers, quoin, lintel, fixing and coursing blocks. Some manufacturers make quarter, half and three-quarter length blocks.

Handling, storage and protection
Unload blocks to a dry, level surface and protect from excessive wetting from the ground and rain. Wet blocks should be allowed to dry before use to reduce drying shrinkage in the completed blockwork. *(see also Section 1.3 'Handling, storage and protection of materials')*

General

Do not mix different block types in the same wall runs. Do not use bricks as closers as they will reduce the insulation value.

Setting out

As a trial, set out the first course dry with 450 mm between the centres of nominal 10 mm cross joints. With so few joints there is little scope for adjustments by varying the joint widths, so consider carefully the position of cut blocks required at window reveals (see 'Bonding', below).

Bedding and jointing

Use only specified mortar mixes which will generally be designation (iii) e.g. 1:1:6 cement:lime:sand or 1:5 masonry cement:sand. Designation (iv) will usually be specified for autoclaved aerated blocks which have a low tensile strength and high shrinkage (see Section 4.1 'Mortars'). Solidly fill bed and cross joints. Do not deeply furrow bed joints (fig 6.110).

Gauge

Take care to maintain gauge to correspond with brickwork gauge. Heavy blocks tend to settle, causing ties to slope down to the inner leaf increasing the risk of water penetration.

Plumbing blocks

Blockwork rises quicker than brickwork, causing fresh, soft, bed joints to be squeezed and deformed. As a result blockwork, particularly the dense type, tends to go out of plumb more readily than brickwork.

Do **not** tap blocks sideways to bring them plumb as this tends to open a gap on one side of

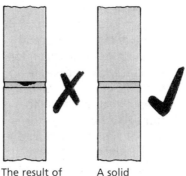

The result of deep furrowing   A solid bed joint

Figure 6.110. *Bedding concrete blocks.*

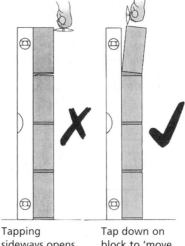

Tapping sideways opens bed joint

Tap down on block to 'move over' plumb and compact bed joint

Figure 6.111. *Plumbing concrete blocks.*

the bed joint causing the block to wobble. Rather than trying to fill this gap with mortar, plumb the block by tapping down on the high side and compact the joint (fig 6.111). Heavier blocks may need to be tapped to plumb and line with a club hammer.

Aligning blocks

Blocks should not be tapped sideways to bring the lower arris into alignment with the work

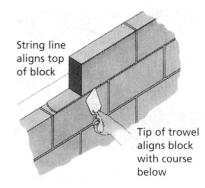

String line aligns top of block

Tip of trowel aligns block with course below

Figure 6.112. *Aligning a concrete block.*

below. This may put the work below out of alignment. Instead, ease the block over the last millimetre or so by using just the tip of the trowel as a lever. The top arris is laid to the string line (fig 6.112).

- In general, cut blocks should be not less than a half-block except, for instance, in every other course at reveals where return or closer blocks may be required (fig 6.113).

Bonding

As blocks are available in many sizes and shapes it is not practicable to illustrate all possible bonding patterns. When setting out a bond pattern for a particular job follow the principles described below.

- Lay blocks to a regular bond pattern, usually half-bond but under no circumstances less than a quarter-block length (fig 6.113).
- Take care that broken bond is no less than quarter bond. If less and particularly if close to a reveal, the vertical line of potential weakness may result in shrinkage cracking (fig 6.115).

Figure 6.113. *Two examples of closing cavities at reveals to openings.*

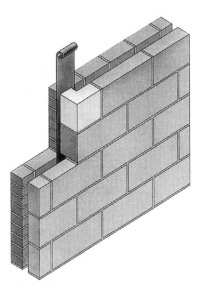

Cut blocks shown in light tone

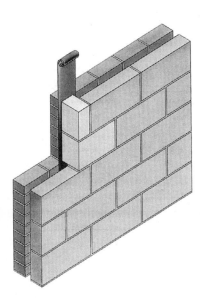

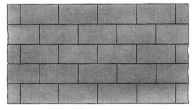

Normal half-bond

Minimum bond lap = ¼ block length

Quarter-bond

Figure 6.114. *Normal and minimum bond pattern.*

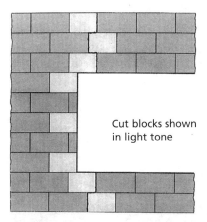

Cut blocks shown in light tone

Figure 6.115. *Badly placed broken bond creating a vertical plane of weakness liable to cracking.*

- Corners may be bonded by cutting standard blocks or by using quoin blocks *(fig 6.116)*.

- 'T'-junctions are normally bonded *(fig 6.117)* Alternatively ties may be used with a straight joint *(fig 6.118)*. Use butterfly ties where vertical differential movement may occur, e.g. where an external leaf is built off a foundation and the partition walls off a suspended floor.

Where vertical movement joints are required to allow horizontal movement, use sleeved debonding ties to allow movement but provide lateral stability. *(see figs 4.39a–c; Section 4.5 'Vertical movement joints')*

Lintels
- Set out bonding so that lintels bear preferably on one whole block *(fig 6.119)*. A minimum two-thirds length block is acceptable but not less *(fig 6.120)*.
- Lintels should normally bear by at least 150 mm.
- Some types of hollow and cellular blocks may need to be filled under lintel ends to provide sufficient bearing strength.

Support during construction
In windy weather walls are readily blown over if not restrained by temporary propping or by fixing floor or flat roof joists. Alternatively the day work lift height may be reduced to suit circumstances.

Hollow and cellular blocks
These can be cut satisfactorily only with a masonry bench saw *(see Section 6.9 'Bricklaying tools and equipment')*. Cellular blocks are laid on a normal mortar bed with the closed end uppermost. Hollow blocks are laid using a shell bedding technique by which the mortar is spread along the outer and inner bed surfaces only.

Movement joints
Since movement characteristics between different types of

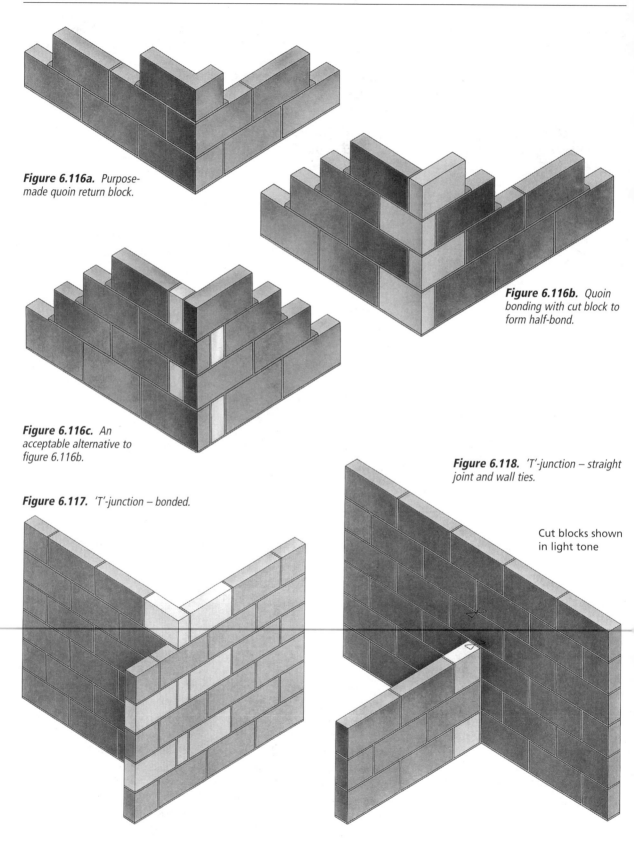

Figure 6.116a. *Purpose-made quoin return block.*

Figure 6.116b. *Quoin bonding with cut block to form half-bond.*

Figure 6.116c. *An acceptable alternative to figure 6.116b.*

Figure 6.118. *'T'-junction – straight joint and wall ties.*

Cut blocks shown in light tone

Figure 6.117. *'T'-junction – bonded.*

Lintel bearing minimum 150 mm

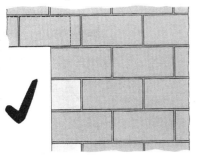

Figure 6.119. *Lintels should preferably bear on whole blocks.*

Lintel bearing minimum 150 mm

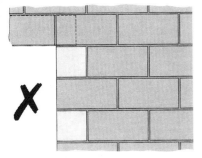

Figure 6.120. *Lintels should not bear on cut block.*

concrete blocks vary widely, recommendations for the spacing of vertical movement joints may differ. But, in general, the spacing should be no more than 6 m in accordance with the Masonry Code of Practice[3] without consulting the manufacturer. Most manufacturers do not consider movement joints necessary in the inner leaves of housing.

References
(1) BS 6073:1981:Part 1 'Specifications for precast concrete masonry units'.
(2) BS 6073:1981:Part 2 'Method for specifying precast concrete masonry units'.
(3) BS 5628:Part 3:1985 'Code of Practice for use of masonry. Materials and components, design and workmanship'.

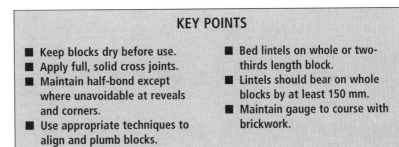

KEY POINTS

- Keep blocks dry before use.
- Apply full, solid cross joints.
- Maintain half-bond except where unavoidable at reveals and corners.
- Use appropriate techniques to align and plumb blocks.

- Bed lintels on whole or two-thirds length block.
- Lintels should bear on whole blocks by at least 150 mm.
- Maintain gauge to course with brickwork.

INDEX